# Future Farming: Advancing Agriculture with Artificial Intelligence

Edited by

**Praveen Kumar Shukla**
*Department of Computer Science & Engineering,*
*Babu Banarasi Das University,*
*Lucknow, UP,*
*India*

&

**Tushar Kanti Bera**
*Department of Electrical Engineering,*
*National Institute of Technology,*
*Drugapur,*
*India*

**Future Farming: Advancing Agriculture with Artificial Intelligence**

Editors: Praveen Kumar Shukla & Tushar Kanti Bera

ISBN (Online): 978-981-5124-72-9

ISBN (Print): 978-981-5124-73-6

ISBN (Paperback): 978-981-5124-74-3

Published by Bentham Science Publishers Pte. Ltd. Singapore. All Rights Reserved.

First published in 2023.

need for a court order if at any point you breach any terms of this License Agreement. In no event will any delay or failure by Bentham Science Publishers in enforcing your compliance with this License Agreement constitute a waiver of any of its rights.

3. You acknowledge that you have read this License Agreement, and agree to be bound by its terms and conditions. To the extent that any other terms and conditions presented on any website of Bentham Science Publishers conflict with, or are inconsistent with, the terms and conditions set out in this License Agreement, you acknowledge that the terms and conditions set out in this License Agreement shall prevail.

**Bentham Science Publishers Pte. Ltd.**
80 Robinson Road #02-00
Singapore 068898
Singapore
Email: subscriptions@benthamscience.net

**BENTHAM SCIENCE**

# CONTENTS

# FOREWORD

It gives me immense pleasure to introduce the book *"Future Farming: Advancing Agriculture with Artificial Intelligence"* edited by Dr. Praveen Kumar Shukla, Department of Computer Science & Engineering, School of Engineering, Babu Banarasi Das University, Lucknow, India and Dr. Tushar Kanti Bera, Department of Electrical Engineering, National Institute of Technology, Durgapur, India. I appreciate the sincere efforts of the editors and authors of the book in framing the content on the most recent and promising area of smart agriculture.

The application of technology-driven solutions in the implementation of agricultural activities is a key concern in the agriculture sector nowadays. Basically, it is one of the most effective ways of raising agricultural production and subsequently fulfilling the food requirements in society. *Artificial Intelligence, the Internet of Things (IoT), Robotics, Drones, Sensors, Telecommunications, Data Analytics, Satellite, Geo-positioning Systems, Unmanned Aerial Vehicles, etc.* are the most promising technologies being utilized in smart agriculture. The implementation of smart agriculture is helping farmers by automating agricultural activities, *i.e.* monitoring of crop health, prediction of crop yield, variable rate irrigation, precise use of fertilizers, diagnosis and curing the crop diseases.

Although there are many advantages of applying technology in the agricultural activities, but its adoption in society is still facing problems. The reasons may be of social acceptance and lack of awareness, lack of light weight intelligent algorithms for precise decision making, data and quality of data for real time decision making, financial issues, security concerns for the deployed systems, *etc*.

I feel that this book will be an outstanding contribution to the researchers working in the area of smart and precision agriculture.

**Vrijendra Singh**
Department of Information Technology,
Indian Institute of Information Technology Allahabad Prayagraj,
India

# PREFACE

Agriculture is one of the oldest professions in the world for feeding the global population. Due to the growth of population and reduction in agricultural land, it is to be planned to move towards smart farming with the help of technology. Artificial Intelligence is playing a vital role in implementing smart methods for agriculture and it will change the aspects of performing agricultural activities. The objective of this book is to explore the applications of artificial intelligence in improving agricultural activities.

The book addresses several aspects of artificial intelligence applications in smart agriculture including, pest control, leaf disease identification, identification of weeds, field security, and applications of drones in smart farming.

A pest control and leaf disease identification system using machine learning technique is implemented where a novel algorithm is proposed, titled Black Window Optimization Algorithm with MayFly Optimization Algorithm (BWO-MA). Also, intelligent recognition and classification of Tomato Leaf Diseases using the Transfer Learning Model are discussed. A pre-trained Squeeze Net Model is used to implement the Transfer Learning Model. Another leaf disease detection system for Millet Leaves is elaborated on in the book. The proposed approach is implemented using Convolutional Neural Network.

Robots are important components in the implementation of smart farming and are playing a big role. A solar-powered robot for the identification of weeds and damage in vegetables is developed and is one of the prime works published in the book. The proposed approach provides the capability for effective control of weeds and damage to crops and also assists in harvesting.

Apart from machine learning and robotics systems, the Internet of Things (IoT) systems are also playing a vital role in the implementation of smart farming and precision agriculture. An IoT-based system powered with AI classification technique is developed for the security of the field and it is mentioned and explained in the book.

Weather conditions are very important for the agricultural system. The prediction of weather conditions will play a big role in planning agricultural activities and dealing with adverse weather conditions. A weather forecasting system for smart farming is also developed using machine learning techniques.

The book also addresses the role of artificial intelligence and drones in smart farming along with the introduction of precision farming, intelligent crop planning and climate smart agriculture.

We hope that the book will surely help the people working in the area of smart agriculture and precision agriculture as it addresses many real-world problems of the agriculture sector through machine learning, IoT, deep learning, *etc*.

**Praveen Kumar Shukla**
Department of Computer Science & Engineering,
Babu Banarasi Das University,
Lucknow, UP, India

&

**Tushar Kanti Bera**
Department of Electrical Engineering,
National Institute of Technology,
Drugapur, India

# List of Contributors

**Avadhesh Kumar**
School of Computer Science & Engineering, Galgotias University, Buddha International Circuit, Greater Noida, Uttar Pradesh, India

**Ajeet Kumar**
Mott Macdonald, Bangalore, India

**Kamatchi Chandrasekar**
Department of Biotechnology, The Oxford College of Science, Chennai, India

**Dahlia Sam**
Department of Information Technology, Dr. MGR Educational and Research Institute, Chennai, India

**Dwijendra Nath Dwivedi**
Krakow university of Economics, Rakowicka, Kraków, Poland

**Harshita Jain**
Geetanjali Institute of Technical Studies, Udaipur, Rajasthan, India

**Gujjula Jhansi**
Department of EEE, Dr. MGR Educational and Research Institute, Chennai, India

**Kalaivani Anbarasan**
Department of Computer Science and Engineering, Saveetha School of Engineering, Saveetha, Institute of Medical & Technical Sciences, Chennai, Tamil Nadu 602105, India

**Kyathanahalli Basavanthappa Vedamurthy**
Karnataka Veterinary, Animal and Fisheries Sciences University, BIDAR, Karnataka, India

**Kitty Tripathi**
Department of Electrical Engineering, Babu Banarasi Das Northern India Institute of Technology, Lucknow, India

**Krishna Kumar Ramaraj**
Department of EEE, School of Engineering, Vels Institute of Science, Technology and Advanced Studies, Chennai, India

**Latif Khan**
Geetanjali Institute of Technical Studies, Udaipur, Rajasthan, India

**Manjunath Chikkamath**
Bosch Global Software Technologies, Bengaluru, India

**Mayank Patel**
Geetanjali Institute of Technical Studies, Udaipur, Rajasthan, India

**Nalinashini Ganesamoorthi**
Department of EIE, R. M. D. Engineering College, Chennai, India

**Nallamilli Pushpa Ganga Bhavani**
Department of Electronics and Communications Engineering, Saveetha School of Engineering, Chennai, Tamil Nadu, India

**Prabhash Chandra Pathak**
School of Computer Applications, Babu Banarasi Das University, Lucknow, India

**Praveen Kumar Shukla**
Artificial Intelligence Research Center, Department of Computer Science & Engineering, School of Engineering, Babu Banarasi Das University, Lucknow, India

**B. Rengammal Sankari**
EEE Department, Dr. MGR Educational and Research Institute, Chennai, India

**Rajan Prasad**
Artificial Intelligence Research Center, Department of Computer Science & Engineering, School of Engineering, Babu Banarasi Das University, Lucknow, India

**Rajashekharappa Thimmappa**
University of Agricultural Sciences, Bangalore, Karnataka, India

**Sujatha Kesavan**
EEE Department, Dr. MGR Educational and Research Institute, Chennai, India

**Santosh Kumar Upadhyay**   School of Computer Science & Engineering, Galgotias University, Buddha International Circuit, Greater Noida, Uttar Pradesh, India

**Saurabh Srivastava**   Geetanjali Institute of Technical Studies, Udaipur, Rajasthan, India

**Syed Anas Ansar**   School of Computer Applications, Babu Banarasi Das University, Lucknow, India

**Sushant Bhatt**   Shri Ramswaroop Memorial University, Lucknow, India

**Srividhya Veerabathran**   Department of EEE, Meenakshi Engineering College, Chennai, Tamil Nadu 600078, India

**Tamilselvi Chandrasekharan**   Department of Information Technology, Dr. MGR Educational and Research Institute, Chennai, India

**Vani Agrawal**   Department of Computer Science and Application, School of Engineering and Technology, ITM University, Gwalior, India

# Enhanced Machine Learning Techniques for Pest Control and Leaf Disease Identification

**Sujatha Kesavan[1,*], Kalaivani Anbarasan[2], Tamilselvi Chandrasekharan[3], Dahlia Sam[3], Nalinashini Ganesamoorthi[4], Kamatchi Chandrasekar[5], Krishna Kumar Ramaraj[6], Nallamilli Pushpa Ganga Bhavani[7], Srividhya Veerabathran[8], B. Rengammal Sankari[1] and Gujjula Jhansi[9]**

[1] *EEE Department, Dr. MGR Educational and Research Institute, Chennai, India*

[2] *Department of Computer Science and Engineering, Saveetha School of Engineering, Saveetha Institute of Medical & Technical Sciences Chennai, Tamil Nadu 602105, India*

[3] *Department of Information Technology, Dr. MGR Educational and Research Institute, Chennai, India*

[4] *Department of EIE, R. M. D. Engineering College, Chennai, India*

[5] *Department of Biotechnology, The Oxford College of Science, Chennai, India*

[6] *Department of EEE, School of Engineering, Vels Institute of Science, Technology and Advanced Studies, Chennai, India*

[7] *Department of Electronics and Communications Engineering, Saveetha School of Engineering Chennai, Tamil Nadu, India*

[8] *Department of EEE, Meenakshi Engineering College Chennai, Tamil Nadu 600078, India*

[9] *Department of EEE, Dr. MGR Educational and Research Institute, Chennai, India*

**Abstract:** The agricultural sector has become an important income source for our country. In terms of nutrient absorption, plant diseases affecting the agricultural yield are creating a great hazard. In agriculture, recognizing infectious plants seems challenging due to the premise of the needed infrastructure. To prevent the spread of diseases, the identification of infectious leaves in the plant is observed to be a necessary step. This work aims to propose a machine learning technique on the ANN method for plant diseases identification and classification. This paper proposes a novel hybrid algorithm, called Black Widow Optimization Algorithm with Mayfly Optimization Algorithm (BWO-MA), for solving global optimization problems.

In this paper, a BWO-MA with Artificial Neural Networks (ANN) based diagnostic model for earlier diagnosis of plant diseases is developed. Comparison has been done with existing machine learning methods with the proposed BWO-MA-based ANN architecture to accommodate greater performance. The comprehensive analysis showed that our proposal achieved splendid state-of-the-art performance.

\* **Corresponding author Sujatha Kesavan:** EEEE Department, Dr. MGR Educational and Research Institute, Chennai, India; E-mail: sujathak73586@gmail.com

**Praveen Kumar Shukla & Tushar Kanti Bera (Eds.)**

**Keywords:** Artificial Neural Networks (ANN), Hybrid black widow optimization algorithm with mayfly optimization algorithm (BWO-MA), Improved canny algorithm, Median filtering, Plant disease.

## INTRODUCTION

The agricultural industry has a vital contribution to global food security and provides a significant amount of nutrients to the population. Besides its ecological significance, the value of farming has expanded across the world, for both human foods and as a tourist attraction. For instance, consumption per capita increased from 9.0 kg in 1961 to 20.2 kg in 2015 [1 - 4]. Agricultural products are one of the most important nutritious foods for humans, as they are high in proteins, vitamins, and minerals, and low in fat. Vitamins A, C, D, K, and Vitamin B2, omega-3 fatty acids, calcium, phosphorus, and minerals such as zinc, iodine, iron, *etc.* are all abundant in them. As a result, they can be considered as a potential remedy to various human health issues. In recent times, agricultural demand is continuing to increase as the world's population grows and the advantages of agricultural products as a source have become more widely acknowledged. The global agricultural production in 2018 was 179.2 million tonnes. Human consumption accounts for around 156 million tonnes of the world's total, which equates to an average of 19.5 kilograms per person. Furthermore, agriculture accounted for 51.5% of the total amount consumed by humans, or 80.3 million tonnes, retaining the remaining for the agricultural industry. Consumption has demonstrated a strong trend of demand in both developed and developing countries in recent years [5 - 11]. The plant species are identified based on their specimens and these specimens identified are based on visual features such as texture, shape, head shape, and color.

Meanwhile, the accurate identification of various species supports scientific research such as ecology, evolutionary studies, plant medicine, and taxonomy [7, 8]. As the agriculture business expands, several concerns develop from current techniques, including the constant occurrence of infectious illnesses in farms, as well as environmental issues that limit agricultural output [1]. Plant illness is one of the most serious risks to many farms; therefore, the use of quick techniques for effective diagnosis is required in addition to experience and knowledge of plant health. Some of the common sources of infection are bacterial, parasitic, fungal infections, and viral infections. Furthermore, the frequency and severity of infections were enhanced by a combination of stressful farm conditions caused by high stocking rates and worsening environmental variables [9 - 13]. Many incurable diseases need the use of professional diagnosis specialists to properly diagnose and treat them. Some plant infections are particularly contagious and spread fast. If the diagnosis is delayed and proper treatment is not administered promptly, the agricultural products will become contaminated and become unusable in a short amount of time. A real-time and remote diagnostic expert system for plant illness is built based on contemporary internet communication technology to accomplish plant disease diagnosis and treatment on time, therefore reducing the risk of damage created to plants [3]. As observation and information technology become advanced, more and more photographs were taken [2]. There are various techniques to detect various plant diseases based on parameters such as visible external signs, atmospheric conditions, and symptoms, behavior signs, water conditions, a captured image of the infected plant, microscopic images, and others. Changes in viewpoint, lighting conditions, and occlusion, among other things, are all issues related to the plant's illness detection. However, image-based tracking and detection are more important for the early illness and recognition of a complex pattern in decision-making of a plant's illness [10]. Pattern matching, physical and statistical behavior, and feature extraction are the essential components of automated plant disease recognition. Plant disease identification is also crucial for plant species counts, population assessments, plant counting, plant association studies, and ecosystem monitoring [7]. Usually, monitoring is done visually in the field or by analyzing photos recorded at crucial points, which necessitates specialized training in addition to being a laborious, time-consuming, and expensive operation. Automatic solutions based on computer vision have been developed to help in the identification of plant species to address these challenges [5, 6].

## RELATED WORK

Jing Hu *et al.* [14] have presented the multi-class support vector machine (MSVM) approach for categorizing plant species through texture features and color. Here, for the classification process, MSVM based one-against-one algorithm was employed. Likewise, Md. Shoaib Ahmed *et al.* [14, 15] have developed the Support Vector Machine (SVM) approach to recognize disease-affected plants. The working process was divided into two phases. The initial phase includes the denoising progression. In the second phase, the classification process was conducted through the SVM with a kernel function. On the other hand, Meng-Che Chuang *et al.* [16, 17] have presented the error-resilient classifier and unsupervised feature learning approaches to recognize plant diseases. Furthermore, an unsupervised clustering approach was utilized for the presented classifier. Likewise, Anderson Aparecido dos Santos, *et al.* [17] have presented the Convolutional Neural Network (CNN) to identify the Pantanal plant species. Similarly, Sourav Kumar Bhoi *et al.* [18, 19] have presented the fuzzy logic-based method and Triangular Membership Function (TMFN) to recognize them as Leaf spots, Leaf Blights, Rusts, Powdery Mildew, Downy Mildew, and Black spot disease in plants. Further, a canny edge detector was utilized to process the black spots in the diseased plant images. On the other hand, Valentin Lyubchenko *et al.* [19] have presented the color image segmentation approach to identify plant disease. Here, the plant surface was taken as the primary information factor. Likewise, Shaveta Malik *et al.* have employed the FAST (Features from Accelerated Segment Test) and Histograms of Oriented Gradients Feature Descriptor to detect diseases in plants.

## BACKGROUND STUDY

### Artificial Neural Network (ANN)

ANN is a computational model that is designed in a way that the human brain analyses and processes information. It is based on Artificial Intelligent (AI) and connects various processing elements; each is similar to a single node. ANN consists of interconnected processing components which are called neurons. All nodes take various signals based on the internal weight as an input and produce a single output. The generated output is the input for another neuron. The architecture of ANN is categorized into different layers such as the input layer, various hidden layers, and the output layer. The input layer accepts the input and processes it. The output layer provides the final output. The mathematical function is performed in the hidden layers and it doesn't have any direct interaction with the user program.

ANN adapts its configuration based on the internal or external data that runs over the network during the learning process. ANN has the ability to mitigate the error, and possibility of recalling, and provides high-speed data. Therefore, it is utilized to solve complex problems like prediction and classification. ANN has been applicable in various fields such as prediction, character recognition, and data forecasting. ANN learning can be either supervised or unsupervised. Supervised training is one of the common neural network trainings, which is accomplished by providing a set of sample data with the expected outcome from every sample to the neural network. Unsupervised training is almost similar to supervised training the only difference is that it does not provide the expected outcome to the neural network. This unsupervised training occurs when the neural network classifies the input into numerous groups. The ANN architecture is shown in Fig. (**1**).

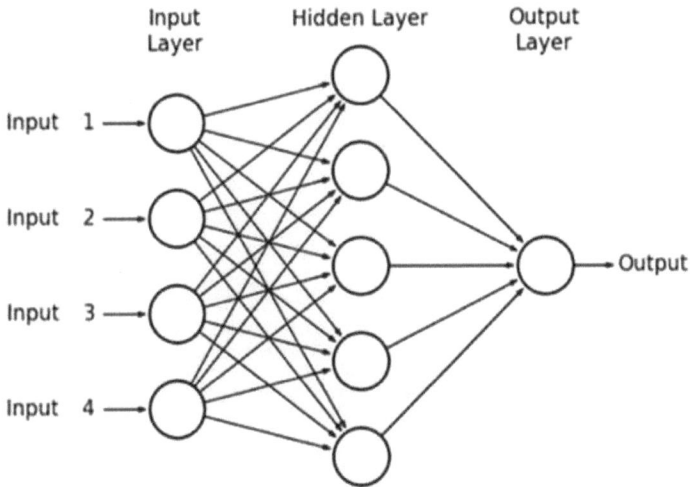

**Fig. (1).** Basic structure of ANN.

## Mayfly Optimization

The main perception of the MFO technique is stimulated by the behavior of mayflies, especially during the mating process. The mayfly belongs to the Ephemeroptera order which is one kind of primitive group of insects called Balaenoptera. The name mayfly is derived from the event that they appear mainly in the UK in the month of May. This optimization technique depends on the PSO and comprises the advantage of GA and FA. Adolescent mayflies are visible to the naked eye after completing the hatching. An adult mayfly only exists for one or two days, until achieving the ultimate goal of breeding. The male attracts the female by assembling swarms, a few meters above the water, and performing a nuptial dance. The female fly enters the swarm and mates with a male. This mating process exists for a few seconds. After completing the mating process, the female mayfly drops the eggs onto the surface of the water and the process goes on. The feasible solution to the problem is denoted by each mayfly position in the search. At first, two sets of mayflies are generated randomly which contain the female and male populations. The swarm is gathered by the movement of the male mayfly which indicates the male mayfly's position. Whereas the female mayfly does not gather the swarm but they fly to the male for the breeding process. The selection of parents is done through the male and female populations. Parent selection happens in the same way when a female attracts a male. The selection is either done in a random process or by their fitness value. At last, the finest female breeds with the finest male mayfly and the process goes on.

## Male Mayflie's Movement

The male mayflies gathering in swarm indicate each male mayfly's location based on its own knowledge or by its neighbour flies. Let us consider, the current position of the fly is $a_m^i m$ in the search space at time step $i$. The position of the flies is varied based on the addition of velocity $v_m^{i+1}$ to the current position. This is given in the below equation.

$$a_m^{i+1} = a_m^i + v_m^{i+1} \tag{1}$$

Moreover, the velocity of the male mayfly $m$ is evaluated as follows:

$$v_{mn}^{i+1} = v_{mn}^i + \propto_1 e^{-\beta r_p^2}\left(pbest_{mn} - a_{mn}^i\right) + \propto_2 e^{-\beta r_g^2}\left(gbest_n - a_{mn}^i\right) \tag{2}$$

where $v_{mn}^i$ indicates mayfly velocity, $pbest_m$ is the best position of the mayfly, $\propto_1, \propto_2$ are positive attractive constants which are utilized to scale the contribution of the cognitive and social component. $r_p$ and $r_g$ are the Cartesian distance between the $\alpha_m$ and $pbest_m$ as well as $\alpha_m$ and $gbest$ respectively. These distances are evaluated as follows:

$$\|a_m - A_m\| = \sqrt{\sum_{n=1}^{j}(a_{mn} - A_{mn})^2} \tag{3}$$

where $\alpha_{mn}$ indicates the $n^{th}$ element of mayfly $m$, $A_m$ is equal to the $pbest_m$ and $gbest$.

## Female Mayflie's Movement

The female mayflies fly in the direction of males for the breeding process. Consider $b_m^i$ as the current position of the female mayfly $m$ in the search space at time step $i$. The position of the flies is varied based on the addition of velocity $v_m^{i+1}$ to the current position. Whereas, the attraction process is performed at random. Therefore, this process is modelled as a deterministic process based on the fitness function that the best male will attract the best female, then the second-best male attracts the second-best female, and so on. Subsequently, the minimization problem is taken into account and its velocities are evaluated as follows.

$$v_{mn}^{i+1} = \begin{cases} v_{mn}^i + \propto_2 e^{-\beta r_{mf}^2}\left(a_{mn}^i - b_{mn}^i\right), & if\ f(b_t) > f(a_t) \\ v_{mn}^i + fl * r, & if\ f(b_t) \leq f(a_t) \end{cases} \tag{4}$$

Where, $r_m$ indicates the cartesian distance between the female and male mayfly, $fl$ indicates the coefficient of the random walk, therefore, the value of $r$ ranges from $[-1,1]$. $v_{mn}^i$ is the velocity of female fly $m$ in dimension $n = 1, , \ldots\ldots x$ at time step $i$.

## *Mating of Mayfly*

The mating process of two mayflies is indicated by the crossover operator. In the mating process, the parent is selected from either the male or female population. Whereas, the male attracted the female based on the parent selection process. Normally, the selection process is carried out either at random or based on a fitness function. In the end, the best female breeds with the best male, then the second-best female breeds with the second-best male, and so on. The outcome of the crossover is the two offsprings which are produced as follows:

$$offspring1 = L * M + (1 - L) * F \tag{5}$$

$$offspring2 = L * F + (1 - L) * M \tag{6}$$

Where, $M$ and $F$ indicate the male and female parent, $L$ indicates the random value within a particular range. Whereas the initial velocities of the offspring are fixed as zero.

## BLACK WINDOW OPTIMIZATION

Based on the concept of the black widow spider, BWO (Black widow optimization) has been developed. The black widow mostly has the character of nocturnal, where the female spider remains blind during the day whereas at night time, the female spins the web. Moreover, the female spends most of her life in the same web where it resides initially. Usually, the female spider starts mating, by the continuous process of this, the female leaves certain spots in the net with the liquid called a pheromone to captivate male spiders. The bizarre thing here is that once the mating has been completed, the female one eats the male spider and saves the egg to the egg sack. Once the egg hatches and offspring have been formed and it's ready for sibling cannibalism, then again there is a chance of child cannibalism where the child consumes its mother in case of lacking in fitness.

The first one is sexual-based cannibalism where the female eats her husband after mating with respect to the fitness value. The second is sibling cannibalism where the healthy spider eats its weaker siblings. As per the cannibalism rate, the concept is utilized. The fittest young ones are alive in the population and others are discarded from it. This is called sibling cannibalism. The third is child cannibalism where the child eats its mother in case of the weaker values.

The female spider lays eggs in the sack after eleven days of the gestation period. The offspring then born and the same is ready for sibling cannibalism. In this process, the young strong spider eats their weakest sibling. The Black widow spider lives together for seven days in the maternal web, this time sibling cannibalism often happens. The density-dependent cannibalism decides the size of the population, this might be significant in the black widow populations where the mother even consumes the young spiders in a short period. The remaining living spiders are considered the fittest young spiders. So based on this black spider concept, the BWO (Black Window Optimization) is achieved.

## Mathematical Evaluation

Based on the random initial population of black widows, the black widow optimization process has been started. This type of randomly generated population encompasses the female and male black widows for the offspring formation. The starting population of the black widow is expressed as,

$$X_{N,d} = \left[ x_{1,1} x_{1,2} x_{1,3} \dots\dots x_{1,d} \right] \tag{7}$$

Where the number of the decision variable is $d$, the population of the black widow is $X_{N,d}$, then the upper bound population is $ub$ then the population number is $N$. The $(X_{N,d})$ is useful in minimizing or maximizing the core function and is denoted by effective solution population as follows,

$$Objective\ Function = f(X_{N,d}) \tag{8}$$

In the proposed BWO model, various predefined parameters are specified such as $Q_{pt}$, $Q_e$, $R_p$, $R_E$, $\Omega_{ts}$, $\Omega_{es}$, $\Omega_{er}$ and $\Omega_{sr}$ which are defined in the previous sections. These parameters are used to indicate the upper and lower bounds of $P_e$ and $P_s$.

The upper bound of $P_e$ is $P_{e\,max}$, the lower bound of $P_e$ is $\frac{Q_E - P_{pt}\Omega_{ts}}{\Omega_{es}}$, the upper bound of $P_s$ is $\frac{Q_P - P_{e.max}\Omega_{er}}{\Omega_{es}}$, the upper bound of $P_s$ is $\frac{Q_P - (Q_E - P_{pt}\Omega_{ts})\Omega_{er}/\Omega_{es}}{\Omega_{sr}}$, and the lower bound of $\frac{Q_P - P_{e.max}\Omega_{er}}{\Omega_{sr}}$.

Mutation is the next step in black widow optimization, where the mutation rate is used for the selection of young spiders. A small random value is added to a selected young spider for the mutation process.

For the mutation process, smaller randomly generated values are incorporated with a selected young sibling spider.

$$Z_{k,d} = Y_{k,d} + \alpha \qquad\qquad (9)$$

where $Z_{k,d}$ is the mutated population of black widows, Here, the randomly generated muted value is $\alpha$, then the randomly selected number is $k$ and the younger spider which is selected randomly is $Y_{k,d}$.

## PROPOSED METHODOLOGY

In terms of nutrient secureness, plant diseases in agriculture yield a greater hazard. However, plant diseases are considered a serious problem among farmers as it tends to spread quickly through the air, water, and soil. For decades, plant diseases have been diagnosed manually by the naked eyes of experienced farmers. In this article, it is needed to discover the disease of the plant that occurred in agriculture. By utilizing the combination of machine learning and clean image processing techniques, we easily discover infectious plants caused by germs. This work has two partitions: The first part is a rudimentary part, in which image pre-processing and segmentation have been implemented to increase the image and mitigate the noise. The feature selection and classification have occurred in the second portion. (Fig. **2**) shows the proposed methodology.

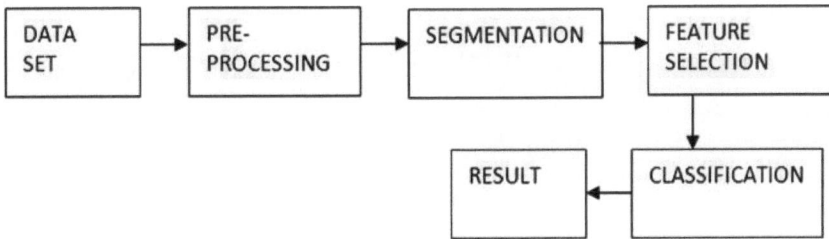

**Fig. (2).** Overall flow of the Proposed Architecture.

The plant disease classification and identification approach have four phases: pre-processing, segmentation, feature selection and classification. The first stage is pre-processing, where the plant images are composed and employed with pre-processing stage, where the filtering approach (median filtering) has been performed. The filtering process usually helps to get rid of unwanted noise from the input image, then to enhance the contrast of the image, the histogram equalization is performed. Once the pre-processing gets completed, by using the improvised canny algorithm, segmentation of the plant images has performed. Once the image segmentation process is completed, the segmentation output undergoes the feature selection process. As a novel contribution to the feature selection, a hybrid algorithm called a Black Widow Optimization Algorithm with

Mayfly Optimization Algorithm (BWO-MA), for solving global optimization problems. The main reason is that the mayflies mutated to enhance the exploration ability of the algorithm. Once the feature selection process is complete, the output is sent to the classification process. A combined (BWO-MA) with Artificial Neural Networks (ANN) based diagnostic model is used for earlier diagnosis of plant diseases. The proposed method is to extract an automatic set of features for classification and identification of plant diseases. The proposed method has high efficiency in improving classification accuracy.

## Pre-processing

It's utilized for pre-processing as it changes the intensity of the image to enhance the contrast of the plant image. Let's assume $In^{im}$ as an input image with the matrix of $J_g$ x $J_s$ integer pixel lies among 0 and 1, and then, $INV$ is defined as the possible count of intensity values. The maximum value taken usually is 256. In eq (10), the NHS histogram of $In^{HE}$ with a bin possible intensity is mentioned. Moreover, the term $he = 0,1 .., INV- 1$ with a bin possible intensity is explained in eq (10).

$$NHS = \frac{number\ of\ pixels\ with\ density\ he}{total\ number\ of\ pixels} \tag{10}$$

$$In^{HE} = floor\left((INV - 1)\sum_{he=0}^{In^{im}} NHS\right) \tag{11}$$

From the above equation, *floor*() usually rounds the values to the closer integer value. As a result, the histogram even makes the $In^{HE}$ plant image into a median filter for the removal of the noise. Though the filter removes the noise, it retains its pixel whether it's high or low but it enhances the edges of the plant images. This median filter is nonlinear. The major idea of the filter is it replaces the noisy pixel with the median value of the nearest pixel which is sorted out based on the grey-level plant images. The result $In^{MF}$ is given based on Equation (13) when the median filter is implemented for the input plant image $In^{HE}$.

$$In^{MF}(a, b) = med\ \{In^{HE}(a - x, b - y)u, x \in H\} \tag{12}$$

The median filtered images and the original filter are denoted by $In^{MF}$, $In^{HE}$ in eq (13) and then $H$ is denoted as the two-dimensional mask. As a result of this, the final pre-processed image is $In^{MF}$. And this step is moved further into the segmentation phase.

## Leaf Image from Plants - Segmentation Model Using Improved Canny Algorithm

Leaf image segmentation using Otsu is a method that creates the separability of the resultant classes maximum to automatically determine the thresholds. Its basic ideas are according to the grey characteristics of leaf images, the image is separated into background and foreground, making their variances of inter-class maximum, finally obtaining the optimal thresholds. For an $M$ X $N$ leaf image $I(x,y)$, the segmented threshold of foreground and background is represented as $T$ the rate of the foreground is $\omega 1$, its average value is $u1$; the rate of background is $\omega 2$, and the average value is $u2$, the mean of the image is $u1$, and the variance of the inter-class is $g1$. Then we can get the following formulas:

$$\omega_1 = \frac{N_1}{M \times N} \tag{13}$$

$$\omega_2 = \frac{N_2}{M \times N} \tag{14}$$

$$N_1 + N_2 = M \times N \tag{15}$$

$$\omega_1 \times \omega_2 = 1 \tag{16}$$

$$u = u_1 \times \omega_1 + u_2 \times \omega_2 \tag{17}$$

$$g = \omega 1 \times (u - u_1)^2 + \omega 2 \times (u - u_2)^2 \tag{18}$$

$$g = \omega 1 \times \omega 2 \times (u_1 - u_2)^2 \tag{19}$$

Finally, by implementing the traversed method which attains the threshold that makes $g$ maximum, and the optimal threshold is $T$. As the highest threshold of $T$, and the low threshold can be obtained as:

$$T_h = k \times T_l \tag{20}$$

Where $k$ is a constant, $k$'s default value is considered to be '2'.

### Steps of Improved Canny Algorithm

Step 1: smooth the image with median filtering

As median filtering approaches smooth the images in such a way that it removes the unnecessary noise from the inputted image where it not only considers the relation of the domain but it takes the relation range by a combination of these two.

Step 2: Compute the gradient magnitude and direction

The magnitudes, direction, and gradient are explained in this, once step (1) gets completed.

Step 3: Perform non-maximum suppression

This step is more similar to the traditional canny operator.

Step 4: Adaptively determine the double thresholds and perform edge detection and connection

Estimate the histogram's magnitude gradient. After that, Otsu algorithm is executed and applied to $T_h$ & T1 $T_h$ & $T_1$ of the doubly created thresholds, then the algorithm scans the entire leaf's image and any marks that are on the leaf for candidate edge points. To perform the segmentation of the leaf, the automatic threshold technique has been utilized. To compute the noisier level in the above segmentation procedure, the interclass among background and foreground objects proceeds.

**Leaf Image Feature Selection Using Hybrid Black Widow Optimization Algorithm with Mayfly Optimization Algorithm (BWO-MA)**

The leaf diseases identification feature selection is performed using a hybrid Black Widow Optimization Algorithm with Mayfly Optimization Algorithm (BWO-MA) and the description of each methodology is given below: As a novel contribution to the feature extraction, a hybrid algorithm called Black Widow Optimization Algorithm with Mayfly Optimization Algorithm (BWO-MA) is used for solving global optimization problems. To improve the exploration capability of the algorithm, the main reason is considered to be the mayflies' transformation which mainly combines the black widow mutation mechanism of the BWO with the mutated behaviors of the MA, thus greatly enhancing the local searching and global searching ability of the algorithm. The following section presents the mathematical models of the initial inhabitants, procreate, cannibalism, transformation, and convergence.

**1. Initial population:** BWO (Black Widow Optimization algorithm), the population is initially executed randomly, where two types of populations are required such as males and females. Based on this initialization, offspring are generated for future generation. In this process, the computation of fitness value is significant; the fitness function is denoted as f at a widow. The following shows a mathematical representation of the black widow spider's initial population.

$$X_{N,d} = \begin{bmatrix} x_{1,1} & x_{1,2} & x_{1,3} \ldots \ldots x_{1,d} \\ x_{2,1} & x_{2,2} & x_{2,3} \ldots \ldots x_{2,d} \\ & x_{N,1} & x_{N,2} \; x_{N,3} \; \ldots \; x_{N,d} \end{bmatrix} \tag{21}$$

$$lb \leq X_i \leq ub$$

Here, $N$ indicates the size of the population, the black spider population is $x_{N,d}$, the problem's number of the decision variable is $N$, then the upper bound and lower bound of the population is $ub, lb$ respectively. The eq (2) represents the effective solution populations $x_{N,d}$ that are useful in reducing or increasing the objective function.

$$\text{RMSE} = \quad \frac{1}{n}\Sigma_{i=1}^{n} w_i (t_i - \hat{t_i})^2 \tag{22}$$

N - No of samples

$t_i$- True sample value

$\hat{t_i}$ - Corresponds to the predictive value

**2. Procreate:** Each pair is independent in the group where they parallelly act for mating to produce a new generation. As discussed, they individually process mating on the web from other spiders. Therefore, in the real-time process, they produce 10K eggs approximately. However, the fittest spider or strongest spider in the web only survives. In this algorithm, an array is considered for the reproduction process, this array-based reproduction is carried out until a widow array with random numbers is available. Then, $\mu$ is denoted for creating an offspring based on the following equation,

x1 and x2 ➔ Parents

y1 and y2 ➔ Offspring

$$y_1 = \mu \times x_1 + (1 - \mu) \times x_2 \tag{23}$$

$$y_2 = \mu \times x_2 + (1 - \mu) \times x_1 \tag{24}$$

i and j can be represented in the range of 1 to N

$\mu$ can be determined in the random range of 0 and 1. cc

**3. Cannibalism:** Cannibalism can be executed by three types such as sexual cannibalism, Child cannibalism, and sibling cannibalism. In sexual cannibalism, the male spider is eaten by the female while mating or after mating. Here, the fitness value is highly considered in this process. The second cannibalism is child eats their parents based on their fitness value to determine whether the spider lings are weak or strong. Likewise, sibling cannibalism is spider eats its sibling if it is weak. In this algorithm, the cannibalism rate is determined for computing the survival rate.

**4. Mutation:** To develop the exploration capability, the newly formed offspring are evolved. As described below, a normally generated random number is added to the variable of the offspring.

$$\text{Offspring'}_n = \text{Offspring}_n + k \tag{25}$$

Where normally distributed random value is *k*.

**5. Convergence:** A three-stop condition could be taken just like the other evolutionary algorithms, which are: i) an already defined total of iterations; ii) for several iterations, there are no changes in the fitness value of the optimal widow which is an observant; iii) when the specific level of accuracy is achieved.

### Pseudo-Code of the Hybrid (BWO-MA) Algorithm

Input: Max- num of iterations, No of cannibalism rate, procreate rate to num. of reproduction is "nr", mutation rate.

### Output: Objective Function's –RMSE

Initialization

The initial population of black widow spiders.

Each pop is a D-dimensional array of chromosomes for a D-dimensional problem.

  a. Fitness value (RMSE) evaluation until termination reached.
  b. Determine nr and find the best solution in pop1(population 1).

Cannibalism & procreating

for i=1 to nr do

from pop1, randomly select two solutions as parents;

By using equation 12, generate D children;

Destroy the father;

Based on the cannibalism rate, destroy some of the children (new achieved solutions);

Save the remaining solutions into pop2;

End for

Mutation

1Offspring mutation

1Replace worst mayflies with the best new offspring created by utilizing equation (25)

Updating

Update the population

Return the best solution from pop;

The obtained best solution is given into the ANN classifier

The performance is evaluated to prove the best classifier.

Stop

## Leaf Image Classification Using (BWO-MA) with ANN

For the wider range of information processing, an ANN (Artificial Neural Network) is utilized which is a softer computing technique. This technique acts as the weighted graph where nodes are considered as the artificial neurons and then the directed edges among neurons are taken as the weights. This network is defined into two types: One is a recurrent network and the other one is a feed-forward network. Recurrent networks are dynamic whereas feed-forward networks are static by nature. This produces output as one set and not in a sequential manner for the provided inputs. As a result of this, the multilayer neural networks were executed successfully in the decision support systems for disease diagnosis Fig. (3).

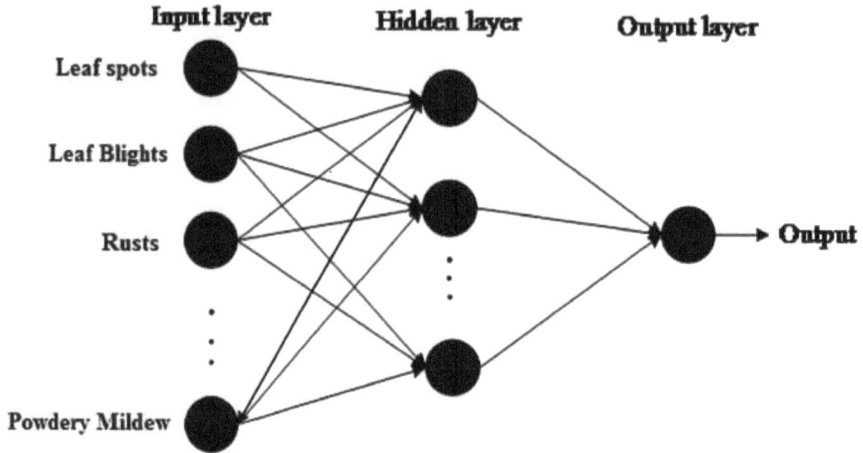

**Fig. (3).** ANN structure for predicting leaf disease.

In MLP (Multilayer Perceptron), there exist multiple layers of neurons, but the three of the minimum layers are the input layer, hidden layer, and one output layer. Here the output layer is responsible for the results. With the nonlinear activation function, every neuron in the other layers acts as the computational component, except from the input layer. The objective of the neural network is that whenever the data exists in the input layer, the neurons are calculated consecutively and then the result is attained at each output neuron. The output of the neural network denotes the suitable classes for the input. Along with weight values, each of the neurons in the hidden layers and the inputs is associated with other neurons of the very next layer. To sum up the threshold and calculate the weighted inputs, the hidden layer's inputs are taken into consideration. The MLP structure of the input, hidden and output layer is demonstrated. The elements of the datasets are shown in the input layer. The hidden layer secures the not linearly separated data, finally, the desired results are provided by the output layer. In the weight function, a threshold node is added in the input layer. By executing the sigmoid function, the result's sums accomplish the activity of the neurons.

In eq. (26) the entire process is explained.

$$p_j = \sum_{i=1}^{n} w_{j,i} \, x_i + \theta_j, m_j = f_j(p_j) \tag{26}$$

Where $p_j$ is assumed to be the linear input combination of $x_1$, $x_2$, ... $x_n$, and $\theta_j$ is considered as the threshold, then $w_{j,i}$ is the connectivity between the neuron 'j' and the input $x_i$, also, the activation function here is $j^{th}$ neuron and $m_j$ is the output. As described in eq (27), sigmoid function is the general choice for activation function.

$$f(t)=1/(1+e^{-t}) \tag{27}$$

For the training of MLP, the back-propagation method has been utilized which is a method of gradient descent method used for the adoption of weights. From the pseudorandom sequence generator, all w (weight vectors) are created with smaller random values. However, this process can take many steps to train the network, then the adjusted weights are calculated at every step. To overcome the above-mentioned problems, a (BWO-MA) based approach is utilized to compute the optimal value of the weight and threshold functions, because (BWO-MA) has the capability to determine weight parallel and finding the optimal solutions.

**Hyper-Parameter Tuning With (BWO-MA)**

Hyper-parameters selection is a significant process of machine learning performance improvement; therefore, the appropriate techniques should be considered for enhancing the tuning process. Thus, the incorporation of effective optimization makes a tremendous impact on the performance of machine learning. Here, a hybrid meta-heuristic algorithm is considered including Black Widow Optimization Algorithm with Mayfly Optimization Algorithm (BWO-MA). Moreover, the BWO-MA approach to solve optimization problems of selecting and tuning hyper-parameters makes a huge difference from other optimization techniques. The enhancement of the exploration ability of the machine learning algorithm was the reason for the selection of the BWO-MA. This algorithm enables a better convergence speed therefore, the performance and results can be obtained as the most effective. As discussed in the previous section, ANN performance is improved by tuning the parameters of ANN with the BWO-MA strategy. One input layer, one hidden layer, and one output layer are the parameters utilized in this work to predict the identification of leaf diseases.

**RESULT AND DISCUSSION**

**Dataset Description**

The leaf dataset (https://www.tensorflow.org/datasets/catalog/plant_village) was utilized for this analysis which includes 6 varieties of leaf species such as Leaf spots, Leaf Blights, Rusts, Powdery Mildew, Downy Mildew, and Black spot, etc.

It contains 915 images with a high resolution, captured by placing the three cameras in different locations. The data set contains 3 visual features. Furthermore, the effectiveness of the system is validated by comparing the evaluation metrics of the proposed approach with the existing approaches. The evaluation metrics are computed by the confusion matrix attained from the experimental outcomes. For this analysis, Accuracy, Precision, Recall, F1_score and specificity are taken as evaluation metrics.

**Accuracy:** Accuracy can be defined as the fraction of the total count of appropriately categorized hand gesture images from the total count of leaf images that are normal and infected. Equation 28 describes the formula to calculate accuracy.

$$\text{Accuracy} = [(T_p+T_n)/(T_p+T_n+F_p+F_n)] \tag{28}$$

**Precision:** Precision can be defined as the ratio of appropriately categorized positive hand gesture images to the total count of positively predicted diseased leaf images. Equation 20 describes the formula to calculate the precision.

$$\text{Precision} = [T_p/(T_p+F_p)] \tag{29}$$

**Recall:** Recall can be defined as the fraction of appropriately categorized positive hand gesture images from the total count of positively predicted diseased leaf images. Equation 21 describes the formula to calculate recall.

$$\text{Recall} = [T_p/(T_p+F_n)] \tag{30}$$

**F1_Score:** F1_Score can be defined as the average harmonic between recall and precision. Equation 22 describes the formula to calculate F1 score.

$$\text{F1score} = [2(\text{Recall x precision})/(\text{Recall} + \text{precision}) \tag{31}$$

**Specificity:** Specificity can be defined as the ratio of the total count of appropriately categorized negative hand gesture images to the total count of negatively predicted diseased leaf images. Equation 32 describes the formula to calculate specificity.

$$\text{Specificity} = [T_n/(T_n+F_p)] \tag{32}$$

The experimental process is carried out in MATLAB 2020-Ra version along with computer window 10 PRO with 8 GB RAM, Intel ® core (TM) i3-6098P CPU @ 3.60 GHz. MATLAB code was utilized to construct and train the ANN framework and also find the hyper-parameters by the hybridization of black widow optimization & mayfly algorithm.

## Evaluation & Results

In this work, the Machine learning technique is incorporated with the meta-heuristic algorithms namely Artificial Neural Network (ANN) with hybrid mayfly and black widow optimization algorithm (BWO-MA) to effectively recognize the hand gesture. The evaluation metrics results have shown better performance of the proposed ANN with BWO-MA approach. Furthermore, different leaf images are taken as an input from the open-source dataset to classify the disease-affected leaves of given images as shown in Fig. (**4**).

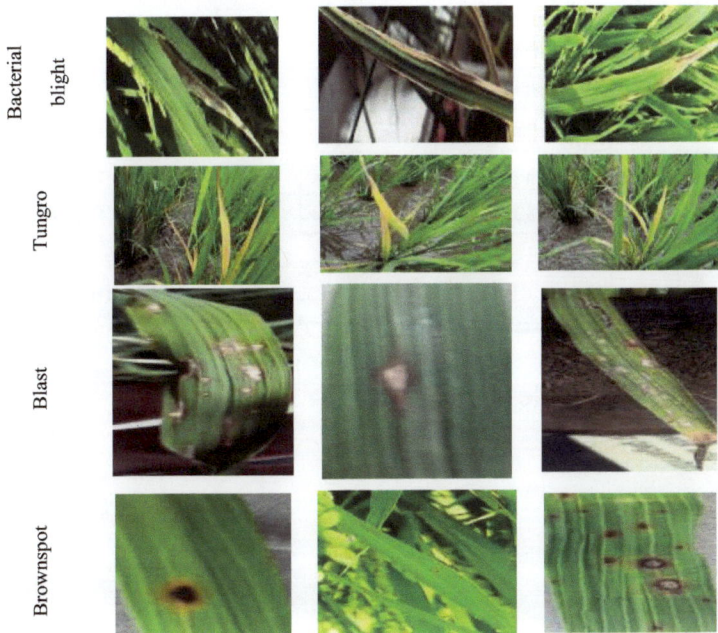

**Fig. (4).** Sample Leaf images affected by various diseases.

The pre-processed images are used as input images. The pre-processing progress is done by applying the median filtering approach. Among all the pre-processing techniques, the median filtering approach gives effective smoothing of spiky noise.

The GLCM was applied to the input sample images. As to extract the features in the feature selection process, the GLCM (Grey level co-occurrence matrix) approaches classified infected regions of the given input images of the leaf. The infected portions are shown by red-colour points.

Table **1** illustrates the evaluation metrics for different approaches. The evaluation metrics for this analysis are accuracy, sensitivity, precision, F1-measure, and specificity. The accuracy for the SVM, Decision tree, Naïve Bayes, and the proposed ANN with BWO-MA approaches is 95.93, 95.43, 97.14, 93.62, and 95.24 respectively. The precision ranges for the SVM, Decision tree, Naïve Bayes, and the proposed ANN with BWO-MA approaches are less as compared with the proposed algorithm, respectively.

**Table 1. Comparative analysis for plant disease detection.**

| Techniques / Performance Metrics | SVM | Decision Tree | Naïve Bayes | Proposed ANN with BWO-MA |
|---|---|---|---|---|
| Accuracy | 92.56 | 83.13 | 77.82 | 95.93 |
| Precision | 88.17 | 84.92 | 82.49 | 95.43 |
| Recall | 90.92 | 78.54 | 77.12 | 97.14 |
| F1_ score | 92.94 | 83.92 | 82.24 | 93.62 |
| Specificity | 86.81 | 80.12 | 78.32 | 95.24 |

In terms of precision, accuracy, F1_score, specificity, and recall, Figure 8 demonstrates better result of the proposed ANN with the BWO-MA approach than the existing previous approaches. Comparing it with all the approaches, Naïve Bayes approach is less effective (Fig.**5**).

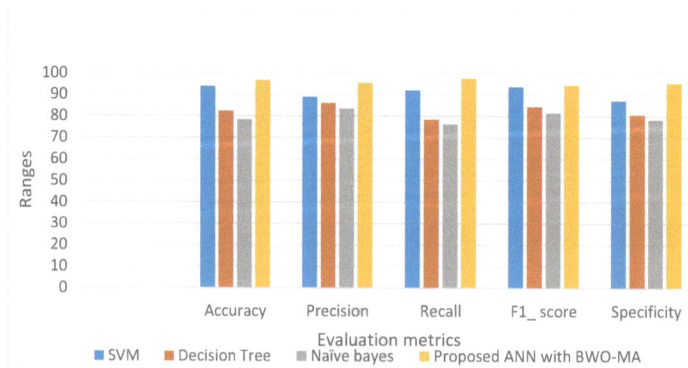

**Fig. (5).** Graphical representation of comparative analysis.

## CONCLUSION

To recognize the infectious leaves, a prominent machine learning-based BWO-MA with ANN was proposed in this work. Along with different models, it showed huge matrices and classified output with visual interaction from the classification outcomes. Also, we implemented image processing techniques such as histogram equalization, canny algorithm segmentation to enhance the classifiers for transforming the input image into a more suitable format with our classifier. In this work, pre-processing, segmentation, hybrid (BWO-MA) feature selection, and ANN classifier approaches are used to enhance the level of prediction and identification on the leaf image. The simulation results demonstrated that the proposed machine learning method is way better than the convolutional machine learning prediction models of leaf image data. This proposed method has high computational efficiency for leaf disease prediction compared to the existing machine learning-based classifiers with high accuracy.

## REFERENCES

[1]    G.T. Abate, T. Bernard, A. de Brauw, and N. Minot, "The impact of the use of new technologies on farmers' wheat yield in Ethiopia: Evidence from a randomized control trial", *Agric. Econ.,* vol. 49, no. 4, pp. 409-421, 2018.
[http://dx.doi.org/10.1111/agec.12425] [PMID: 30166743]

[2]    I. Ali, F. Cawkwell, E. Dwyer, and S. Green, "Modeling managed grassland biomass estimation by using multitemporal remote sensing data A machine learning approach", *IEEE J. Sel. Top. Appl. Earth Obs. Remote Sens.,* vol. 10, no. 7, pp. 3254-3264, 2017.
[http://dx.doi.org/10.1109/JSTARS.2016.2561618]

[3]    J. Amara, B. Bouaziz, A. and Algergawy, "A deep learning-based approach for Banana leaf diseases classification. B. Mitschang *et al.* (Hrsg.): BTW 2017-Workshopband". *Lecture notes in Informatics (LNI), Gesellschaft fÉur Informatik, Bonn*, pp. 79-88.

[4]    S. Amatya, M. Karkee, A. Gongal, and Q. Zhang, "Science Direct detection of cherry tree branches with full foliage in planar architecture for automated sweet-cherry harvesting", *Biosyst. Eng.,* pp. 1-13, 2015.

[5]    B.S. Anami, N.N. Malvade, and S. Palaiah, "Deep learning approach for recognition and classification of yield affecting paddy crop stresses using field images", *Artificial Intelligence in Agriculture,* vol. 4, pp. 12-20, 2020.
[http://dx.doi.org/10.1016/j.aiia.2020.03.001]

[6]    K.R. Aravind, P. Raja, and M. Pérez-Ruiz, "Task-based agricultural mobile robots in arable farming: A review", *Span. J. Agric. Res.,* vol. 15, no. 1, p. e02R01, 2017.
[http://dx.doi.org/10.5424/sjar/2017151-9573]

[7]    S.H. Awan, S. Ahmed, N. Safwan, Z. Najam, M.Z. Hashim, and T. Safdar, "Role of internet of things (IoT) with Blockchain Technology for the Development of Smart Farming", *Journal of Mechanics of Continua and Mathematical Sciences,* vol. 14, pp. 170-188, 2019.

[8]    M. Ayaz, S. Member, and M.A. Uddin, "Internet-of-things (IoT) based smart agriculture: Towards making the fields talk", *IEEE Access,* vol. 7, pp. 129551-12953, 2019.
[http://dx.doi.org/10.1109/ACCESS.2019.2932609]

[9]    V. Badrinarayanan, A. Kendall, and R. Cipolla, "SegNet: A deep convolutional encoderdecoder architecture for image segmentation", *IEEE Trans. Pattern Anal. Mach. Intell.,* vol. 39, no. 12, pp.

2481-2495, 2017.
[http://dx.doi.org/10.1109/TPAMI.2016.2644615] [PMID: 28060704]

[10]   M.D. Bah, E. Dericquebourg, A. Hafiane, and R. Canals, "Canals, Deep Learning Based Classification System for Identifying Weeds Using High-Resolution UAV Imagery", In: *Advances in Intelligent Systems and Computing,* K. Arai, S. Kapoor, R. Bhatia, Eds., vol. 857. Springer: Cham, 2018.

[11]   N. Balakrishnan, and G. Muthukumarasamy, "Crop production—ensemble machine learning model for prediction", *International Journal of Computer Science and Software Engineering,* vol. 5, pp. 148-153, 2016.

[12]   L. Bencini, S. Maddio, G. Collodi, D. Di Palma, G. Manes, and A. Manes, "Development of Wireless Sensor Networks for Agricultural Monitoring", In: *Smart Sensing Technology for Agriculture and Environmental Monitoring. Lecture Notes in Electrical Engineering.,* S. Mukhopadhyay, Ed., vol. 146. Springer: Berlin, Heidelberg, 2012.
[http://dx.doi.org/10.1007/978-3-642-27638-5_9]

[13]   K. Bhagawati, A. Sen, K.K. Shukla, and R. Bhagawat, "Application of data Mining in Agriculture Sector", *International Journal of Computer Science Trends and Technology,* vol. 3, pp. 66-69, 2016.

[14]   B.B. Bhanu, K.R. Rao, J.V.N. Ramesh, and M.A. Hussain, "Agriculture field monitoring and analysis using wireless sensor networks for improving crop production", *IFIP International Conference on Wireless and Optical Communications Networks (WOCN),* 2014.
[http://dx.doi.org/10.1109/WOCN.2014.6923043]

[15]   K. Bodake, R. Ghate, H. Doshi, P. Jadhav, and B. Tarle, "Soil based fertilizer recommendation system using internet of things", *MVP Journal of Engineering Sciences,* vol. 1, 2018.

[16]   R. Bogue, "Robots poised to revolutionise agriculture", *Ind. Rob.,* vol. 2, pp. 468-471, 2016.

[17]   P. Boniecki, K. Koszela, H. Piekarska-Boniecka, J. Weres, M. Zaborowicz, S. Kujawa, A. Majewski, and B. Raba, "Neural identification of selected apple pests", *Comput. Electron. Agric.,* vol. 110, pp. 9-16, 2015.
[http://dx.doi.org/10.1016/j.compag.2014.09.013]

[18]   P. Bosilj, T. Duckett, and G. Cielniak, "Connected attribute morphology for unified vegetation segmentation and classification in precision agriculture", *Comput. Ind.,* vol. 98, pp. 226-240, 2018.
[http://dx.doi.org/10.1016/j.compind.2018.02.003] [PMID: 29997405]

[19]   M. Brahimi, K. Boukhalfa, A. Moussaoui, and M. Brahimi, "Deep learning for tomato diseases: Classification and symptoms visualization deep learning for tomato diseases: Classification and symptoms visualization", *Appl. Artif. Intell.,* vol. 31, no. 4, pp. 299-315, 2017.
[http://dx.doi.org/10.1080/08839514.2017.1315516]

**CHAPTER 2**

# Automatic Recognition and Classification of Tomato Leaf Diseases Using Transfer Learning Model

**Santosh Kumar Upadhyay**[1,*] and **Avadhesh Kumar**[1]

[1] *School of Computer Science & Engineering, Galgotias University, Buddha International Circuit, Greater Noida, Uttar Pradesh, India*

**Abstract:** Timely diagnosis of plant disease is important to get better crop yields. Infected plants can cause significant financial losses to farmers by lowering crop yields. It is extremely desirable to detect early signs and symptoms of plant diseases in a nation like India, where agriculture supports the majority of the population. More accurate and faster plant disease detection might assist in lowering the damage. With tremendous improvements and advancements in deep learning, the effectiveness and precision of plant disease detection and identification systems may be improved. The goal of this study is to discover leaf illnesses found in tomato crops and reduce the financial losses caused by the diseases. We have implemented transfer learning using a pre-trained Squeeze Net Model to detect and classify tomato leaf diseases. Our model can automatically detect 9 types of deadly diseases that are very common in tomato crops. We have acquired 10 classes (one healthy leaf class and 9 diseased leaf classes) consisting of 16,012 tomato leaf samples from a benchmarked Plant Village dataset to train and validate the suggested method. On the public dataset, the class-wise classification precision rate varies from 77.9% to 99.6%, and the overall classification accuracy of the suggested model is observed as 93.1% which is a significant enhancement in performance over previous tomato disease detection techniques.

**Keywords:** CNN, Deep learning, Plant disease, Tomato, Transfer learning, SqueezeNet.

## INTRODUCTION

Tomato is the globe's most common vegetable, and it may be seen in every kitchen, regardless of region and culture. Most people across the globe like tomatoes and their product very much. It is consumed in many different ways.

---

* **Corresponding author Santosh Kumar Upadhyay:** School of Computer Science & Engineering, Galgotias University, Buddha International Circuit, Greater Noida, Uttar Pradesh, India; E-mail: ersk2006@gmail.com

Tomato is the 3rd most often planted crop after potato and sweet potato. India comes at the second position in the world for producing tomatoes. However, due to numerous infections, the quantity and quality of the tomato crop suffer. These losses can be reduced by providing an efficient disease diagnosis system in farming [1]. Plant illnesses are now very difficult to identify using the traditional approach of visual observation by humans. Conventional methods concentrate mostly on professionals' expertise, and manuals [2], but the bulk of them are costly, time-taking, and labour-intensive, and these methods face difficulty to identify exactly [3]. As a result, a speedy and precise method for identifying plant diseases plays a vital role in the welfare of agriculture's business and environment.

Computer vision systems have been advanced to the point where they can provide quick, normalised, and correct solutions to these issues. The adoption of an automated disease diagnosis approach is advantageous in detecting a plant disease in its early stages. Deep learning has recently achieved outstanding results in disciplines such as object detection [4], voice recognition [5], biomedical image classification [6, 7], and object recognition. The deep Convolution Network has shown excellent results in the field of crop disease recognition. In the present study, we have applied CNN based. Among these 9 diseases to diagnose and classify 9 types of tomato diseases. Of these 9 diseases, 5 diseases namely target spot, early blight, late blight, tomato leaf mold, and Septoria leaf spot are caused by fungal infection, 2 diseases namely tomato yellow leaf curl virus and tomato mosaic virus are caused by the virus infection, 1 disease namely bacterial spot is caused by bacterial infection and one disease namely two-spotted spider mite is caused by spider mite. Leaf samples of these tomato diseases are depicted in Fig. (**1**) in section 3.2.1.

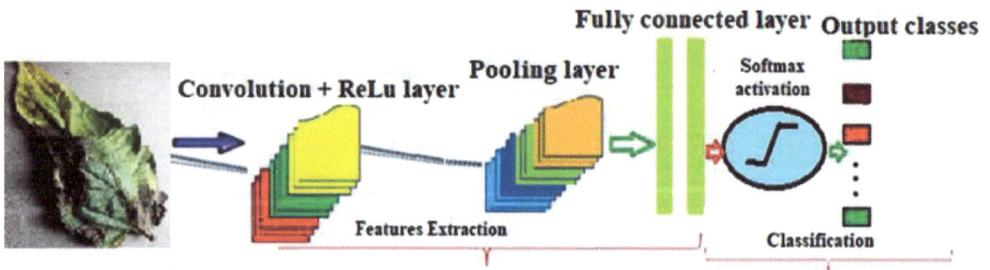

**Fig. (1).** Basic structure of CNN.

The following is a breakdown of the chapter's structure: Sect. 2 discusses related studies, emphasizing key contributions. Sect. 3 discusses the materials and methods. The database of infected leaves is discussed, which covers plant category illnesses types, and the number of image samples in each disease class.

The methods explain how the dataset was processed, how the model was implemented, and how the model was trained. The experimental findings and analysis are presented in Section 4, which includes a comparison of existing methodologies with the suggested method. Finally, Section 5 brings the study to a close and suggests some future enhancements.

## EXISTING WORKS

El-Helly *et al.* [8] employed an artificial neural network to diagnose leaf miner-damaged, downy and powdery mildew-infected leaves in cucumber plants in 2004. Liu *et al.* [9] employed backpropagation neural networks to forecast the incidence of illnesses and pests in apple plants in 2005. The authors have used the preceding eleven years' field information to train the network. Sammany and Medhat [10] employed a genetic algorithm to refine learning parameters and the design of neural networks (NN) used in disease detection. Refinement was done to achieve optimal parameters. These optimized NNs and support vector machines (SVMs) were utilized to detect plant diseases.

Tian *et al.* [11] presented a plant disease recognition system using an SVM classifier for wheat crops. They converted the RGB color format of input leaf images into HIS format, then GLCM was used to find 7 immutable moments as shape features. Finally, the SVM classifier used extracted features to identify wheat leaf diseases. Dhaygude *et al.* [12] utilized Spatial Gray-level Dependence Matrices to get texture information from leaf images. The RGB image is transformed into HSV format to remove the green part from the image. Thus, infected portions are extracted and further segmented into equal-sized patches. Patches covering more than fifty percent of information are further analyzed.

Selvaraj *et al.* [13] proposed an SVM-based plant disease detection system. They have developed a 4 steps procedure: first, a color transformation scheme is constructed for the input RGB image, then the green pixels are masked and changed using various threshold levels, and next, segmentation is performed. Texture statistics are generated for usable segments, and the resultant features are then sent to the support vector machine for classification. Further, Pujari *et al.* [14] introduced a plant disease diagnosis and classification model based on SVM and ANN. They have taken samples of leaf illness caused by fungi in cereals for study. The k-means clustering method was used to separate the infected regions from the leaf. The features related to the texture and color properties of the infected region were retrieved and utilized as inputs for the classification model.

To improve the classification rate, Zhang *et al.* [15] proposed a strategy for identifying cucumber illness based on the breakdown of the global-local singular value. The SVM model was used to classify the illness of leaf images.

Cheng Liu *et al.* [16] examined two strategies for identifying Rice false smut in natural settings. A total of 693 naturalistic environment photos were gathered. The first is to create a classification model that combines SVM classification with HOG feature extraction (Histogram of Oriented Gradient). The second model was built using a new deep convolution structure. To assess the merits of the new system, the authors compared it to Alex Net and VGGNet-11. In the result analysis, the proposed novel CNN architecture has shown a better performance than SVM. Suresha *et al.* [17] proposed a strategy for detecting Brown Spot and Blast illnesses in paddy crops. The diseases were categorized and recognized using the global threshold approach and the kNN classifier. For the suggested technique, a classification accuracy of 76.59 percent was attained.

In 2018, Zhang [18] *et al.* presented a deep learning-based automated detection and classification system for diseases found in maize plants. 500 digital photos consisting of 8 types of diseases were collected from different sources. They have modified Cifar10 and GoogLeNet deep convolutional neural architectures to achieve a high classification rate of 98.8% and 98.9%, respectively.

In 2019, Baranwal *et al.* [19] built a CNN architecture to identify apple leaf diseases. They have collected healthy and unhealthy leaf image samples from the PlantVillage dataset. Image filtering, compression, and generation operations were used to create a large dataset for the training of the suggested model. In the result's analysis, the proposed model achieved an overall accuracy of 98.54%. Further, Shrivastava *et al.* [20] employed deep CNN transfer learning to perform their research. They have taken three types of rice leaves diseases in their study. The suggested transfer learning model achieved a 91.37% classification accuracy.

A convolutional neural network was proposed by Hari *et al.* [21] as a useful method for detecting illnesses in several kinds of plants such as grape, maize, tomato, and apple plant. Total 15210 leaf photos from ten groups were utilised to train and test the model. The accuracy of the proposed convolutional neural network was 86%.

Upadhyay and Kumar [22] proposed a novel method to diagnose rice diseases using Otsu's thresholding and deep convolutional architecture. Otsu's thresholding technique was used to segment the leaf part from the acquired input images. Segmented images were given to the suggested CNN model to recognize and classify the 3 types of rice diseases. The result analysis achieved an accuracy of 99.7%.

Santosh and Avadhesh [23] introduced an effective method to diagnose brown spot disease in paddy crops at an early stage. They have used 2 stage segmentation to locate the infection in input leaf images. Deep CNN architecture was used to detect the brown spot at an early stage. The suggested model performed well with a classification accuracy of 99.20%.

In 2020, Chen *et al.* [24] used the VGGNet convolution neural architecture combined with the Inception module to detect plant diseases. They have applied transfer learning to transfer the weight and biases of the pre-trained VGGNet model on a huge dataset ImageNet to initialize the weight and bias parameters of the suggested model. On the public dataset, the suggested technique obtained a classification accuracy of 91.83 percent, which was a significant enhancement over previous existing related approaches.

Using PlantVillage dataset, Agarwal *et al.* [25] created a Convolution neural architecture with three convolutional, three pooling, and two fully connected layers to recognize one healthy and nine tomato illnesses categories, with an overall accuracy of 91.2 percent. Elhassouny and Smarandache [26] used pre-trained MobileNet to create an application for smartphones to recognize and categorize nine tomato crop illnesses (Howard *et al.*, 2017). The plantVillage dataset was applied to provide diseased leaf image samples for training and validation of the model. An accuracy rate of 90.3% was observed in the result analysis.

Zaki *et al.* [27] utilized pre-trained deep learning architecture MobileNetV2 to implement transfer learning to recognize and classify the leaf diseases in tomato crops and used a CNN model based on transfer learning and fine-tuned MobileNetV2 for tomato leaf disease classification. The findings revealed that the introduced model has a precision rate of 90% in detecting diseases. In 2021, to diagnose rice illnesses, Anandhan and Singh [28] suggested utilizing mask R-CNN and Faster R-CNN. They utilized a smartphone camera to take rice leaf photos, then used 1500 digital photographs in experimentation to provide training to the suggested model. According to the evaluation of the trial findings, Mask R-CNN is found more effective for recognizing different infections namely; sheath blight, brown spot, and blast, with an accuracy of 94.5%, 95%, and 96%, respectively.

## MATERIALS AND METHODS

This section describes the cutting-edge techniques, methods, and data samples utilized to achieve the classification results.

## Related Works

### *Convolution Neural Network*

In recent times, CNN is widely being used in the application of image classification and object detection. The deep convolutional neural network is a robust and efficient model that uses specialized convolution and pooling operations to detect plant diseases using image samples of plant leaves, stems, or fruits. CNN is a classification framework that divides the input pictures into labeled categories. CNN's multiple layers capture visual information prior to applying the classification process to categorize the input images. CNNs use picture data, provide learning to the model, and then automatically categorize features for better classification. Convolution, pooling, and activation layers are the 3 major parts of CNNs.

Image identification, facial recognition, and video analysis are a few common areas where CNNs are employed. A typical CNN structure is shown in Fig. (**1**) and consists of the following important components.

### *Convolution Layer*

CNN's basic component is the convolutional layer. It captures the input signal's high-level characteristics. The convolution layer takes an input picture and transforms it by convolving it with a filter.

This process generates a number of feature maps for an input picture. When a specific-sized filter is applied to a picture, the weighted sum of pixel values corresponding to the filter's values is computed. As a result of this calculation, multiple feature maps for the input data are generated.

### *Activation Layer*

After the convolve operation, the ReLU activation is utilized to convert all negative values to zero. ReLU is recommended because of its ability to quickly converge.

### *Pooling Layer*

After the convolution layer, the pooling layer is applied. Pooling operations are set up based on the requirements. Average pooling, Max-pooling, and min-pooling are three separate pooling operations. The pooling technique is mostly used to reduce dimensionality and to choose the most important feature.

## Fully Connected Layer

The fully connected layer, which includes a SoftMax activation function, receives these extracted properties of images to recognize and categorize the input sample into proper output classes.

## SqueezeNet

In 2016, Iandula *et al.* presented this design [29]. SqueezeNet is a pre-trained CNN model that uses architectural tactics to minimise the number of parameters, particularly through the introduction of fire modules, which "squeeze" parameters employing convolutions with filters of dimensions 1x1. To increase the effectiveness of standard CNN architecture, it employs 3 basic methodologies. To begin with, the bulk of the kernels in the network structure are 1x1 rather than 3x3, which drastically decreases the number of network parameters. Next, the amount of input channels is reduced to kernels with dimensions 3x3. This method further decreases the number of network parameters significantly. Finally, for bigger activation maps, downsampling is done later in the network. The hypothesis is that there is a direct link between the dimension of the activation maps used for downsampling and the final classification performance.

## PROPOSED WORK

This section discusses all the steps required to design the disease classification and recognition model. The flow diagram of the proposed system is depicted in Fig. (**2**).

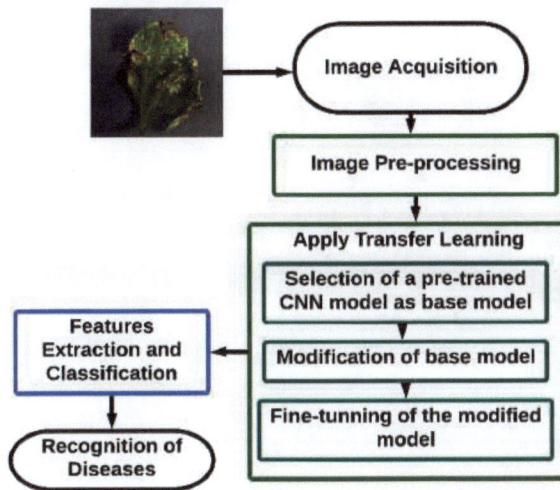

**Fig. (2).** Flow diagram of the proposed system.

## Image Acquisition (Dataset)

To train and validate the suggested technique, we collected 16,012 tomato leaf samples from the benchmarked Plant Village dataset [30] in 10 classes (one healthy leaf class and nine damaged leaf classes). A list of 10 classes along with a number of leaf images is outlined in Table **1**.

**Table 1. Description of diseased and healthy classes.**

| Categories of Tomato Diseases | Training Sample Counts |
|---|---|
| Bacterial Spot | 2,127 |
| Early Blight | 1,000 |
| Late Blight | 1,909 |
| Leaf Mold | 952 |
| Septoria Leaf Spot | 1,771 |
| Two-Spotted Spider Mite | 1,676 |
| Target Spot | 1,404 |
| Yellow Leaf Curl Virus | 3,209 |
| Tomato Mosaic Virus | 373 |
| Healthy Leaves | 1,591 |

The sample images of the 10 classes along with disease names are shown in Fig. (**3**). The whole dataset is partitioned into the proportion of 70:30 for training and validating the suggested model.

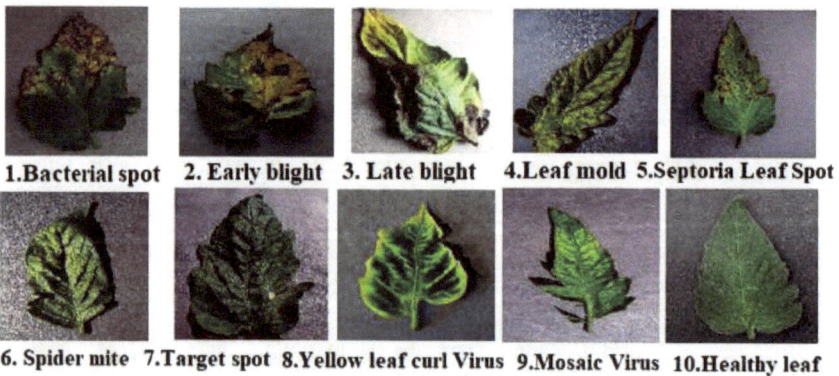

1.Bacterial spot    2. Early blight    3. Late blight    4.Leaf mold    5.Septoria Leaf Spot

6. Spider mite    7.Target spot    8.Yellow leaf curl Virus    9.Mosaic Virus    10.Healthy leaf

**Fig. (3).** Image samples of diseased and healthy leaf.

## Image Pre-Processing

First of all, the acquired input image samples are scaled to generate resized images with dimensions of 227*227 pixels. Then resized images are augmented to get many transformed images before providing training to the model. These transformed images are achieved by slight rotation, scaling, translation, and reflection. These transformed images are used as a large variety of training samples during the training of the model to stop the network from memorizing the data pattern. Therefore, these transformed images reduce the chance of overfitting.

## Establishing a New Deep Network Using Transfer Learning

Transfer learning [31] is an emerging style of learning approach in which deep convolutional architecture that has been trained for one job is utilised as the foundation for a model for a different task. In the traditional learning process, we assign the initial weight of the network with random values. But here, rather than providing the learning from the beginning, we may assign the initial weights of the deep network with the learned weights of the pre-trained model. In this research, we investigate employing pre-trained CNN from the enormous ImageNet typical dataset, which we then apply to the tomato leaf disease samples of the PlantVillage dataset. The following are the major steps of the transfer learning strategy.

***Selection of Base Model of Transfer Learning:*** The base model (Pre-trained CNN) is chosen and the learned weights of this model are used to initialize the weights of the deep network. In our study, we have used SqueezeNet pre-trained CNN model as the base model of transfer learning.

***Modification of the Base Deep Network:*** The weights of the bottom layers of the deep network are frozen. The structure of the remaining top layers can be changed by adding more layers, deleting the layers, and replacing the layers to get a newly updated deep network architecture. In this study, we have frozen the bottom 63 layers of pre-trained SqueezeNet and the remaining top 5 layers are modified. Customization of the top 5 layers is shown in Fig. (**4**).

***Fine-Tuning of Modified Deep Network:*** Labelled tomato leaf images acquired from PlantVillage dataset are used to fine-tune the newly built deep network to reduce the loss function.

## Recognition and Classification

The pre-trained CNN model SqueezeNet is customized so that it can fit on our particular dataset. In this direction, the last learnable convolution layer is replaced with a convolution layer in which the number of filters is kept the same as the number of classes (in our case 10) in the tomato leaf dataset. This replacement is shown in Fig. (**4**). SoftMax layer uses a SoftMax activation function to compute class-wise decimal probabilities. This probability is used to decide the appropriate output class for the given input image sample.

| Layer Number | Native Layers | Modified layers |
|---|---|---|
| 64 | Convolution2dLayer  (1000 convolutions with activation 14x14x1000) | Convolution2dLayer (10 convolutions   with activation 14x14x10) |
| 65 | Relu_convolution (with   activation 14x14x1000) | Relu_convolution (with   activation 14x14x10) |
| 66 | Average   pooling (with activation 1x1x1000) | Average   pooling (with activation 1x1x10) |
| 67 | Prediction SoftMax (with activation 1x1x1000) | Prediction SoftMax (with activation 1x1x10) |
| 68 | Classification Layer (1000 classes) | Classification Layer (10 output classes) |

**Fig. (4).** Customization of SqueezeNet.

## EXPERIMENTAL RESULTS AND DISCUSSION

### Experimental Setting and Environment

The experiment was carried out on a Windows 10 Laptop equipped with a Core i5 processor@ 2.40 GHz, 8GB RAM and 512 GB SSD. MATLAB R2019a was used to implement the deep learning network model. Experimental results were evaluated using MATLAB R2019a. Model training and evaluation were done using diseased and healthy tomato leaf samples acquired from the PlantVillage dataset. 70% of images of the dataset were utilized to train the model. 30% of images of the dataset were used to validate the model. Class-wise distribution of a number of training and testing image samples is shown in Table **2**.

**Table 2. Disease-wise partition of training and validation samples.**

| Categories of Tomato Diseases | Training Sample Counts | Validation Sample Counts |
|---|---|---|
| Bacterial Spot | 1,489 | 638 |
| Early Blight | 700 | 300 |
| Late Blight | 1,336 | 573 |
| Leaf Mold | 666 | 286 |
| Septoria Leaf Spot | 1,240 | 531 |
| Two-Spotted Spider Mite | 1,173 | 503 |
| Target Spot | 983 | 421 |
| Yellow Leaf Curl Virus | 2,246 | 963 |
| Tomato Mosaic Virus | 261 | 112 |
| Healthy Leaves | 1,114 | 477 |

## Evaluation Metrics

The performance of the suggested deep learning model is evaluated in terms of Overall Accuracy, Precision, and Recall parameters. These evaluation metrics are computed in terms of true-positive (TP), True-Negative (TN), false-positive (FP), and false-negative (FN). The overall accuracy of the model can be computed using Equ. 1, the classification Precision of each class can be computed using Equ. 2, and the classification Recall of each class can be computed using Equ. 3.

$$Overall\ Accuracy \tag{1}$$
$$= (Sum\ of\ True\ Positives\ of\ all\ classes)/Number\ of\ classes \tag{}$$
$$Precision = TP/(TP + FP) \tag{2}$$
$$Recall = TP/(TP + FN) \tag{3}$$

## Experiment Deployment and Result Analysis

The experiment was carried out with a tomato diseased leaf dataset, which is acquired from the PlantVillage dataset with one healthy leaves class and nine different types of tomato diseased leaf classes. The accuracy of the training and validation process over the iteration number is shown in Fig. (**5**). The loss rate of the training and validation process is shown in Fig. (**6**).

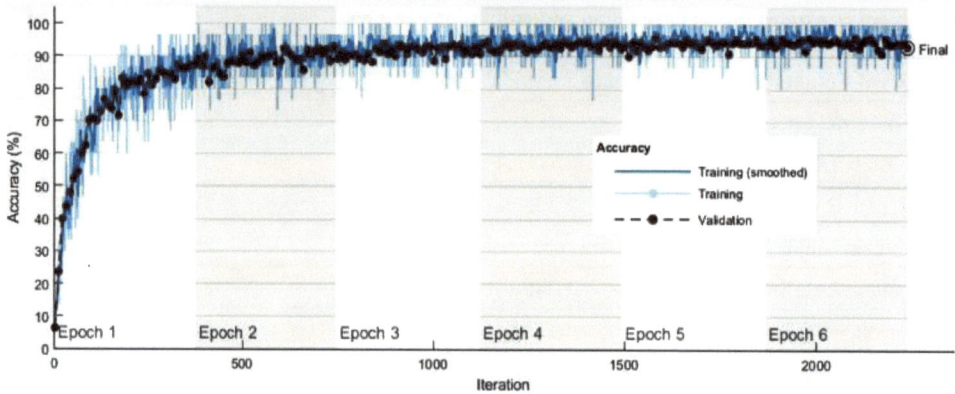

**Fig. (5).** Training and validation progress of the model.

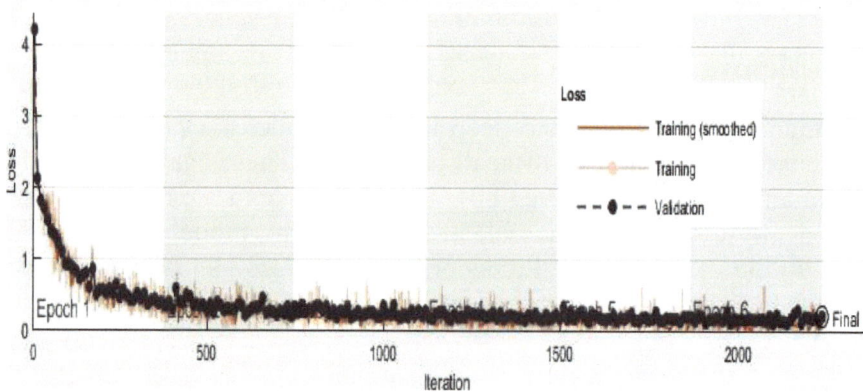

**Fig. (6).** Training and Validation loss rate of the model.

We used 6 epochs to complete the 2238 iterations. Every epoch has 373 iterations to evaluate the learning performance of the suggested transfer learning model.

The accurate and false prediction details of the suggested model are shown in the confusion matrix illustrated in Fig. (7). It is demonstrated that the model learned the features in such a way that false predictions are reduced to zero in the healthy leaves class, while 26 instances of Bacterial spot out of 638, 50 instances of tomato Late blight out of 573, 2 instances of Septoria Leaf Spot out of 531, 39 instances of Target Spot out of 421, 17 instances of Yellow Leaf Curl Virus out of 963, and 1 instance of Mosaic Virus out of 112 were misclassified by the model. Thus, the true positive rates (Recall) of healthy, Bacterial spot, Late blight, Septoria Leaf Spot, Target Spot, Yellow Leaf Curl Virus, and Mosaic Virus class were observed as 100%, 95.9%, 91.3%, 99.6%, 90.7%, 98.2%, and 99.1% respectively.

**Confusion Matrix**

| Output Class \ Target Class | Bacterial Spot | Early Blight | Late Blight | Leaf Mold | Septoria Leaf Spot | Two-Spotted Spider Mite | Target Spot | Yellow Leaf Curl Virus | Tomato Mosaic Virus | Healthy Leaves | |
|---|---|---|---|---|---|---|---|---|---|---|---|
| Bacterial Spot | 612 / 12.7% | 4 / 0.1% | 1 / 0.0% | 0 / 0.0% | 1 / 0.0% | 0 / 0.0% | 0 / 0.0% | 15 / 0.3% | 0 / 0.0% | 0 / 0.0% | 96.7% / 3.3% |
| Early Blight | 1 / 0.0% | 218 / 4.5% | 6 / 0.1% | 0 / 0.0% | 0 / 0.0% | 0 / 0.0% | 0 / 0.0% | 0 / 0.0% | 0 / 0.0% | 0 / 0.0% | 96.9% / 3.1% |
| Late Blight | 2 / 0.0% | 15 / 0.3% | 523 / 10.9% | 0 / 0.0% | 0 / 0.0% | 0 / 0.0% | 0 / 0.0% | 0 / 0.0% | 0 / 0.0% | 0 / 0.0% | 96.9% / 3.1% |
| Leaf Mold | 0 / 0.0% | 0 / 0.0% | 0 / 0.0% | 228 / 4.7% | 0 / 0.0% | 1 / 0.0% | 0 / 0.0% | 0 / 0.0% | 0 / 0.0% | 0 / 0.0% | 99.6% / 0.4% |
| Septoria Leaf Spot | 4 / 0.1% | 45 / 0.9% | 31 / 0.6% | 43 / 0.9% | 529 / 11.0% | 0 / 0.0% | 26 / 0.5% | 0 / 0.0% | 1 / 0.0% | 0 / 0.0% | 77.9% / 22.1% |
| Two-Spotted Spider Mite | 0 / 0.0% | 1 / 0.0% | 0 / 0.0% | 4 / 0.1% | 0 / 0.0% | 446 / 9.3% | 1 / 0.0% | 2 / 0.0% | 0 / 0.0% | 0 / 0.0% | 98.2% / 1.8% |
| Target Spot | 19 / 0.4% | 13 / 0.3% | 2 / 0.0% | 1 / 0.0% | 0 / 0.0% | 36 / 0.7% | 382 / 8.0% | 0 / 0.0% | 0 / 0.0% | 0 / 0.0% | 94.3% / 15.7% |
| Yellow Leaf Curl Virus | 0 / 0.0% | 2 / 0.0% | 2 / 0.0% | 0 / 0.0% | 0 / 0.0% | 1 / 0.0% | 0 / 0.0% | 945 / 19.7% | 0 / 0.0% | 0 / 0.0% | 99.5% / 0.5% |
| Tomato Mosaic Virus | 0 / 0.0% | 2 / 0.0% | 0 / 0.0% | 9 / 0.2% | 1 / 0.0% | 9 / 0.2% | 2 / 0.0% | 0 / 0.0% | 111 / 2.3% | 0 / 0.0% | 82.8% / 17.2% |
| Healthy Leaves | 0 / 0.0% | 0 / 0.0% | 8 / 0.2% | 1 / 0.0% | 0 / 0.0% | 10 / 0.2% | 10 / 0.2% | 0 / 0.0% | 0 / 0.0% | 477 / 9.9% | 94.3% / 5.7% |
| | 95.9% / 4.1% | 72.7% / 27.3% | 91.3% / 8.7% | 79.7% / 20.3% | 99.6% / 0.4% | 88.7% / 11.3% | 90.7% / 9.3% | 98.2% / 1.8% | 99.1% / 0.9% | 100% / 0.0% | 93.1% / 6.9% |

**Fig. (7).** Classification Results

Only 3 diseased classes namely Early Blight, Leaf Mold, and Two-Spotted Spider Mite were classified with slightly low true positive rates as 72.7%, 79.7%, and 88.7% respectively. This low true positive rate is due to the high resemblance of Early Blight as Septoria Leaf Spot, Leaf Mold as Septoria Leaf Spot, and Two-Spotted Spider Mite as Target Spot.

The overall accuracy of the suggested transfer learning model is achieved as 93.10% is illustrated in the bottom-rightmost column of the confusion matrix shown in Fig. (7).

Class-wise Recall (true positive rates) of the proposed transfer learning model is shown in Fig. (8). Class-wise precision of the proposed transfer learning model is illustrated in Fig. (9).

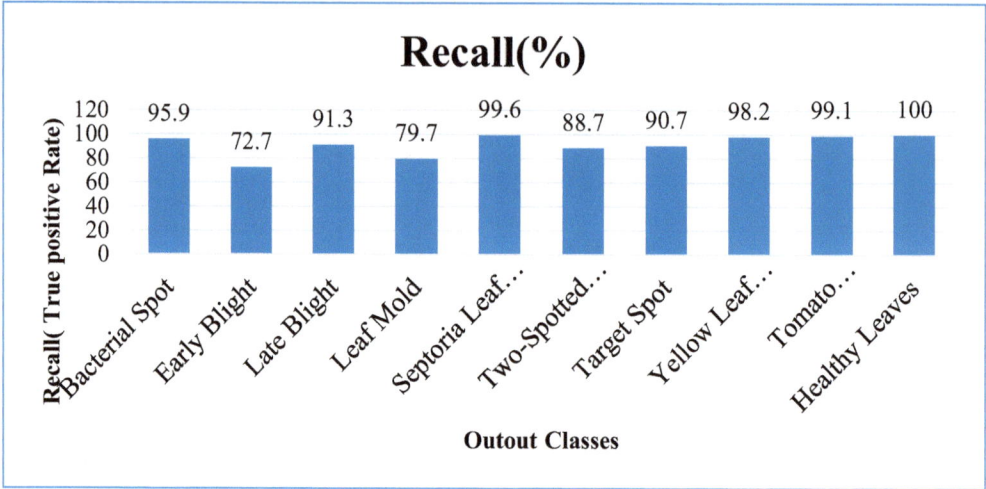

**Fig. (8).**  True positive rates of the classification result.

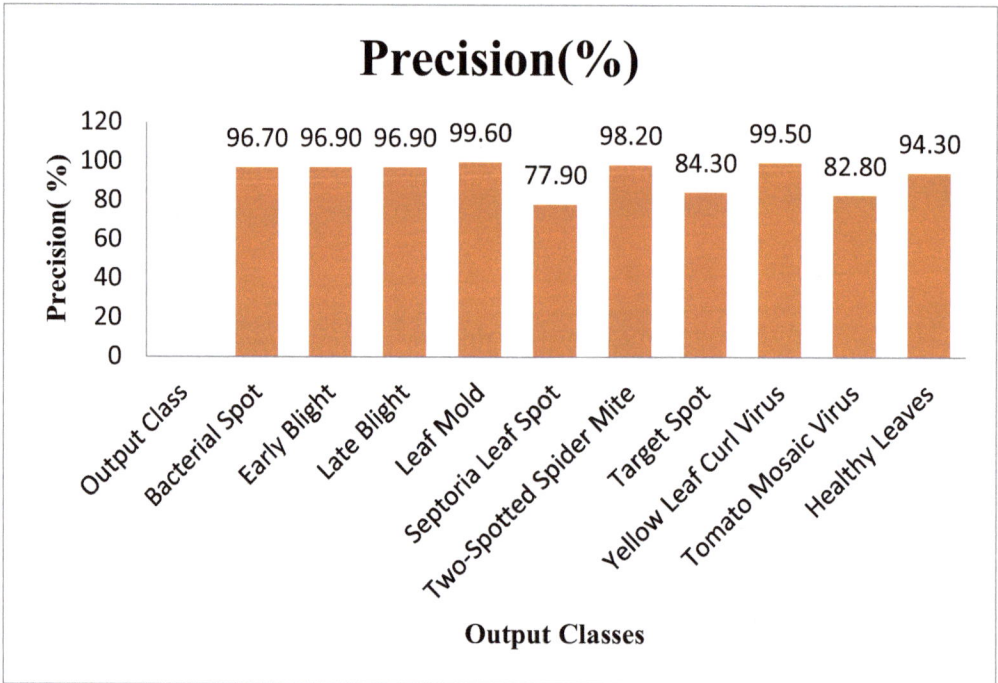

**Fig. (9).**  Class wise precision of classification result.

## Comparison with Earlier Works

We have done a comparative analysis of our approach with 3 related existing works. All the compared methods used leaf images of tomato diseases as input that was acquired from the PlantVillage dataset. Methods used in this comparison utilized 9 same types of diseases for study. Two approaches Elhassouny and Smarandache, 2019 [26], Zaki *et al.*,2020 [27] among these existing works utilized a transfer learning approach based on pre-trained CNN models to recognize the tomato diseases whereas the remaining one approach Agarwal *et al.*, 2020 [25] utilized simple convolution neural architecture with 3 convolutional, 3 pooling, and 2 fully connected layers to recognize the diseases. Our suggested approach outperforms these compared methods in terms of accuracy. The accuracy comparison of the proposed system with earlier works is outlined in (Table **3**).

**Table. 3 Accuracy comparison of existing works and suggested model**

| Authors | Methods | Plant | No. of Diseases | Overall Accuracy |
|---|---|---|---|---|
| Agarwal *et al.*, 2020 [25] | Simple CNN | Tomato | 9 | 91.20% |
| Elhassouny and Smarandache, 2019 [26] | MobileNet | Tomato | 9 | 90.30% |
| Zaki *et al.*,2020 [27] | MobileNetV2 | Tomato | 9 | 90.00% |
| Proposed approach | SqueezeNet | Tomato | 9 | **93.1%** |

## CONCLUSION AND FUTURE SCOPE

The SqueezeNet based deep transfer learning approach was introduced in this book chapter to categorize tomato plant diseases based on their leaf shape. Using automatic feature extraction and shape features, this method can classify 9 different types of tomato diseases. We used the SqueezeNet deep CNN model pre-trained on the huge labeled dataset (ImageNet) to initialize the weights to the suggested deep learning model, rather than performing the training from scratch with random initialization of the weights. The suggested technique achieves an average accuracy of 93.10% for the disease classification of the tomato plant. This method performed better with leaf images even under difficult background characteristics.

For future work, different image segmentation approaches can be used to remove the background from the sample images and extract the diseased region to classify the diseases more accurately.

## ACKNOWLEDGEMENTS

The authors thank Galgotias University, Greater Noida and their family members for their valuable support and they also thank reviewers for their valuable comments.

# REFERENCES

[1]     G. Dhingra, V. Kumar, and H.D. Joshi, "Study of digital image processing techniques for leaf disease detection and classification", *Multimedia Tools Appl.,* vol. 77, no. 15, pp. 19951-20000, 2018.
        [http://dx.doi.org/10.1007/s11042-017-5445-8]

[2]     V. Singh, and A.K. Misra, "Detection of plant leaf diseases using image segmentation and soft computing techniques", *Inf. Process. Agric.,* vol. 4, no. 1, pp. 41-49, 2017.
        [http://dx.doi.org/10.1016/j.inpa.2016.10.005]

[3]     S.P. Mohanty, D.P. Hughes, and M. Salathé, "Using deep learning for image-based plant disease detection", *Front. Plant Sci.,* vol. 7, p. 1419, 2016.
        [http://dx.doi.org/10.3389/fpls.2016.01419] [PMID: 27713752]

[4]     J. Redmon, S. Divvala, R. Girshick, and A. Farhadi, "You only look once: Unified, real-time object detection", *Proceedings of the IEEE Conference on Computer Vision and Pattern Recognition,* pp. 779-788, 2016.
        [http://dx.doi.org/10.1109/CVPR.2016.91]

[5]     O. Abdel-Hamid, A. Mohamed, H. Jiang, L. Deng, G. Penn, and D. Yu, "Convolutional neural networks for speech recognition", *IEEE/ACM Trans. Audio Speech Lang. Process.,* vol. 22, no. 10, pp. 1533-1545, 2014.
        [http://dx.doi.org/10.1109/TASLP.2014.2339736]

[6]     M. Gour, and S. Jain, "Stacked convolutional neural network for diagnosis of covid-19 disease from x-ray images", *arXiv,* 2020.

[7]     M. Gour, S. Jain, and T. Sunil Kumar, "Residual learning based CNN for breast cancer histopathological image classification", *Int. J. Imaging Syst. Technol.,* vol. 30, no. 3, pp. 621-635, 2020.
        [http://dx.doi.org/10.1002/ima.22403]

[8]     M. El-Helly, S. El-Beltagy, and A. Rafea, "Image analysis-based interface for diagnostic expert systems", *Proceedings of the Winter International Symposium on Information and Communication Technologies,* pp. 1-6, 2004.

[9]     G. Liu, H. Shen, X. Yang, and Y. Ge, "Research on prediction about fruit tree diseases and insect pests based on neural network", In: *IFIP—The International Federation for Information Processing.* vol. Vol. 187. Springer: Berlin, Germany, 2005, pp. 731-740.
        [http://dx.doi.org/10.1007/0-387-29295-0_79]

[10]    M. Sammany, and T. Medhat, "Dimensionality reduction using rough set approach for two neural networks-based applications", In: *Rough Sets and Intelligent Systems Paradigms.* Springer: Berlin, Germany, 2007, pp. 639-647.
        [http://dx.doi.org/10.1007/978-3-540-73451-2_67]

[11]    U. Tian, C. Zhao, S. Lu, and X. Guo, "SVM-based multiple classifier system for recognition of wheat leaf diseases", *Proceedings of 2010 Conference on Dependable Computing (CDC'2010).*

[12]    S. B. Dhaygude, and N. P. Kumbhar, "Agricultural plant leaf disease detection using image processing", *International Journal of Advanced Research in Electrical, Electronics and Instrumentation Engineering,* vol. 2, no. 1, 2013.

[13]    A. Selvaraj, N. Shebiah, S. Ananthi, and S.V. Varthini, "Detection of unhealthy region of plant leaves and classification of plant leaf diseases using texture features", *Agric. Eng. Int. CIGR,* vol. 15, no. 1, pp. 211-217, 2013.

[14]    J. Pujari, R. Yakkundimath, and A. Byadgi, "Classification of fungal disease symptoms affected on cereals using color texture features", *International Journal of Signal Processing, Image Processing and Pattern Recognition,* vol. 6, no. 6, pp. 321-330, 2013.
        [http://dx.doi.org/10.14257/ijsip.2013.6.6.29]

[15]   S. Zhang, and Z. Wang, "Cucumber disease recognition based on Global-Local Singular value decomposition", *Neurocomputing,* vol. 205, pp. 341-348, 2016.
[http://dx.doi.org/10.1016/j.neucom.2016.04.034]

[16]   C. Liu, C. Xu, S. Liu, D. Xu, and X. Yu, "Study on identification of Rice False Smut based on CNN in a natural environment", *2017 10th International Congress on Image and Signal Processing, BioMedical Engineering and Informatics (CISP-BMEI),* pp. 1-5, 2017.
[http://dx.doi.org/10.1109/CISP-BMEI.2017.8302016]

[17]   M. Suresha, K.N. Shreekanth, and B.V. Thirumalesh, "Recognition of diseases in paddy leave using KNN classifier", *2017 2nd International Conference for Convergence in Technology (I2CT),* pp. 663-666, 2017.
[http://dx.doi.org/10.1109/I2CT.2017.8226213]

[18]   X. Zhang, Y. Qiao, F. Meng, C. Fan, and M. Zhang, "Identification of Maize Leaf Diseases Using Improved Deep Convolutional Neural Networks", *IEEE Access,* vol. 6, pp. 30370-30377, 2018.
[http://dx.doi.org/10.1109/ACCESS.2018.2844405]

[19]   S. Baranwal, S. Khandelwal, and A. Arora, "Deep learning convolutional neural network for apple leaves disease detection", *Proceedings of International Conference on Sustainable Computing in Science, Technology and Management (SUSCOM)*, Amity University Rajasthan, Jaipur - India, February 26-28, 2019. 20. V. K. Shrivastava, M. K. Pradhan, S. Minz, and M. P. Thakur, "Rice plant disease classification using transfer learning of deep convolution neural network", *Int Arch Photogramm Remote Sens Spatial Inf Sci*, vol. 42-3, pp. 631–635, 2019.

[20]   V. K. Shrivastava, M. K. Pradhan, S. Minz, and M. P. Thakur, "Rice plant disease classification using transfer learning of deep convolution neural network", *Int Arch Photogramm Remote Sens Spatial Inf Sci*, vol. 42-3, pp. 631–635, 2019.

[21]   S.S. Hari, M. Sivakumar, P. Renuga, and S. Suriya, "Detection of plant disease by leaf image using convolutional neural network", *2019 IEEE International Conference on Vision towards Emerging Trends in Communication and Networking (ViTECoN),* pp. 1-5, 2019.
[http://dx.doi.org/10.1109/ViTECoN.2019.8899748]

[22]   S.K. Upadhyay, and A. Kumar, "A novel approach for rice plant diseases classification with deep convolutional neural network", *International Journal of Information Technology,* vol. 14, no. 1, pp. 185-199, 2022.
[http://dx.doi.org/10.1007/s41870-021-00817-5]

[23]   S.K. Upadhyay, and A. Kumar, "Early-stage brown spot disease recognition in paddy using image processing and deep learning techniques", *TS Traitement Signal,* vol. 38, no. 6, pp. 1755-1766, 2021.
[http://dx.doi.org/10.18280/ts.380619]

[24]   J. Chen, J. Chen, D. Zhang, Y. Sun, and Y.A. Nanehkaran, "Using deep transfer learning for image-based plant disease identification", *Computers and Electronics in Agriculture,* vol. Vol. 173, 2020.

[25]   M. Agarwal, A. Singh, S. Arjaria, A. Sinha, and S. Gupta, "ToLeD: Tomato leaf disease detection using convolution neural network", *Procedia Comput. Sci.,* vol. 167, pp. 293-301, 2020.
[http://dx.doi.org/10.1016/j.procs.2020.03.225]

[26]   A. Elhassouny, and F. Smarandache, "Smart mobile application to recognize tomato leaf diseases using convolutional neural networks", *2019 International Conference of Computer Science and Renewable Energies (ICCSRE),* pp. 1-4, 2019.
[http://dx.doi.org/10.1109/ICCSRE.2019.8807737]

[27]   S.Z.M. Zaki, M. Asyraf Zulkifley, M. Mohd Stofa, N.A.M. Kamari, and N. Ayuni Mohamed, "Classification of tomato leaf diseases using MobileNet v2", *IAES International Journal of Artificial Intelligence (IJ-AI),* vol. 9, no. 2, pp. 290-296, 2020.
[http://dx.doi.org/10.11591/ijai.v9.i2.pp290-296]

[28] A. K and A. S. Singh, "Detection of Paddy Crops Diseases and Early Diagnosis Using Faster Regional Convolutional Neural Networks", *2021 International Conference on Advance Computing and Innovative Technologies in Engineering (ICACITE),* pp. 898-902, 2021.

[29] F.N. Iandola, S. Han, M.W. Moskewicz, K. Ashraf, W.J. Dally, and K. Keutzer, "SqueezeNet: AlexNet-level accuracy with 50x fewer parameters and <0.5 MB model size", *arXiv,* 2016.

[30] Availaible from: https://www.kaggle.com/emmarex/plantdisease (accessed on 10 Nov 2021).

[31] A. Lumini, and L. Nanni, "Deep learning and transfer learning features for plankton classification", *Ecol. Inform.,* vol. 51, pp. 33-43, 2019.
[http://dx.doi.org/10.1016/j.ecoinf.2019.02.007]

## CHAPTER 3

# Detection and Categorization of Diseases in Pearl Millet Leaves using Novel Convolutional Neural Network Model

**Manjunath Chikkamath[1,\*], Dwijendra Nath Dwivedi[2], Rajashekharappa Thimmappa[3] and Kyathanahalli Basavanthappa Vedamurthy[4]**

[1] *Bosch Global Software Technologies, Bengaluru, India*

[2] *Krakow university of Economics, Rakowicka 27, Kraków, 31-510, Poland*

[3] *University of Agricultural Sciences, Bangalore, Karnataka, India*

[4] *Karnataka Veterinary, Animal and Fisheries Sciences University, BIDAR, Karnataka, India*

**Abstract:** Pearl millet is a staple food crop in areas with drought, low soil fertility, and higher temperatures. Fifty percent is the share of pearl millet in global millet production. Numerous types of diseases like Blast, Rust, Bacterial blight, *etc.*, are targeting the leaves of the pearl millet crop at an alarming rate, resulting in reduced yield and poor production quality. Every disease could have distinctive remedies, so, wrong detection can result in incorrect corrective actions. Automatic detection of crop fitness with the use of images enables taking well-timed action to improve yield and in the meantime bring down input charges. Deep learning techniques, especially convolutional neural networks (CNN), have made huge progress in image processing these days. CNNs have been used in identifying and classifying different diseases across many crops. We lack any such work in the pearl millet crop. So, to detect pearl millet crop diseases with great confidence, we used CNN to construct a model in this paper. Neural network models use automatic function retrieval to help in classify the input image into the respective disease classes. Our model outcomes are very encouraging, as we realized an accuracy of 98.08% by classifying images of pearl millet leaves into two different categories namely: Rust and Blast.

**Keywords:** Convolutional neural network, Deep learning, Machine learning, Plant disease detection, Pearl Millet Leaves.

## INTRODUCTION

Pearl millet is a staple food crop in regions characterized by drought, low soil fertility and extreme temperatures. Greater than 90 million desperately terrible

\* **Corresponding author Manjunath Chikkamath:** Bosch Global Software Technologies, Bengaluru, India;
E-mail: dr.chikkamath@gmail.com

**Praveen Kumar Shukla & Tushar Kanti Bera (Eds.)**

individuals who live inside the drier components of Africa and Asia, locations where other crops just won't grow, rely on pearl millet for their revenue and daily diet. Fortunately, pearl millet is not only an irrepressible and reliable basis of energy in these areas, but also a better resource for other food requirements, particularly micronutrients. It is cultivated over 26 million hectares in some of the toughest semi-arid tropical environments. It is also eaten as food and feed for livestock. It is the 6th largest cereal crop in the world after maize, rice, wheat, barley, and sorghum. One of the biggest threats to pearl millet production to the farmer is the irreparable crop damage caused due to plant diseases. Millet diseases that affect crops can reduce the quality and quantity of the product by reducing productivity or by completely destroying the crop. Plant diseases found with bare eyes are not always accurate, and in areas where pearl millet is cultivated, finding trained workers or plant pathologists is very difficult. Early detection and identification of plant disease are often impossible in different regions of the earth because of insufficient trained manpower and plant pathologists.

The leaves of the plants initiate the process of photosynthesis whereby the plants obtain their energy. Illnesses/disorders alter the leaves of crops to such an extent that they do not supply adequate food for healthy crop growth, thereby causing poor health or even crop demise. The Food and Agriculture Organization predicts that diseases can cause damage to up to 10 to 30 percent of the world's food production, which is a bigger threat to achieving food security. Thus, manual monitoring of plant diseases is very challenging owing to its complicated nature and lengthy procedure. The traditional method of disease diagnosis is to physically examine specimens by qualified peasants or trained workers or plant pathologists. However, it does take a long time and consequently, it can impact the crop production. In some cases, owing to geographical limitations and inadequate skills of farmers, the plant disease cannot be identified properly [1]. Traditional techniques involve a great deal of time to investigate diseases, while at a similar point in time, crops can suffer from more harm due to real diseases. In addition, situations have arisen where, because of insufficient disease knowledge or wrong interpretation of the severity of the disease, an excessive or inadequate dose of the insecticide has caused severe crop loss. Consequently, it is necessary to reduce manual effort, at the same time making correct predictions and making sure that the lives of farmers are hassle-free. A lot of work has been carried out on the automatic identification of diseases for various crops using image processing techniques [2 - 6]. But not a good amount of work has been carried out in pearl millet disease identification with novel techniques like image processing [7]. So, automatic disease identification using the image has the potential to address all the above issues by automatically detecting and classifying diseases. The novel image processing techniques such as computer vision, and deep learning methods can be valuable in identifying plant disease. From the year 2012, Deep Neural

Networks (DNNs), and especially CNNs have been extremely popular in numerous computer vision tasks, like image detection, identification, and categorization [8].

The goal of this study is to construct a model using CNN to automatically identify the diseases of pearl millet crops from images. This document is organized as follows: Section 2 deals with relevant and significant work in this area. Section 3 explains about the data, data pre-processing and methodology used to build the model and the measures taken to achieve the desired results. In Section 4, we have discussed the results obtained. Section 5 contains the concluding remarks about our work and highlights the opportunity for future work.

## LITERATURE STUDY AND RELATED WORK

A new light CNN with channel mixing operation and multiple size module (L-CSMS) for identifying the severity of plant diseases is proposed [9]. Advanced feature extraction techniques for several crop categories are used [9 - 11]. A method for automatic detection of crop diseases using deep ensemble neural networks (DENN) is also proposed [9]. The performance of DENN surpasses advanced pre-formed algorithms like ResNet 50 and 101, InceptionV3, DenseNet 121 and 201, MobileNetV3 and NasNet. A two-step CNN model to identify crop diseases and classify citrus diseases using leaf images has been developed [12]. The model provided 94.37% precision in citrus fruit disease detection with a 95.8% precision score. The AlexNet algorithm is to rapidly and accurately identify corn diseases [13]. Using various iterations such as 25, 50, 75 and 100, the model achieved 99.16% precision. An adaptive snake segmentation model has also been developed to segment and identify infected areas [14]. Adaptive snake segmentation model is a 2-stage model, *i.e.*, common segmentation and absolute segmentation. A threshold is also proposed based on adaptive intensity for the automated segmentation of powdery mildew disease, making this technique invariant for image quality and stochasticity [15]. The proposed technique has been validated on the complete cherry leaf image series with 99% accuracy. An automatic system, computer-based method of identifying yellow disease, also known as chlorosis is proposed in a study [16] which is a major pulse crop such as Vigna mungo.

A process of optimizing Henry gas solubility (MHGSO) based on mutation to optimize the hyperparameters of the DenseNet-121 architecture is introduced in a study [17]. When tested with a field dataset with complex backgrounds, the MHGSO-optimised DenseNet-121 architecture achieves accuracy and recall scores of 98.81%, 98.60% and 98.75%, respectively. The Philodendron leaf was obtained from a natural grayscale color and tinting, saturation and value technique were applied to the gray image for disease recognition [18]. Also, a computer vision system for classifying medicinal leaves at the corresponding maturity level is presented [19].

The computer vision framework presented by CNN can provide a classification precision of about 99% for the simultaneous prediction of foliar species and maturity stage. A deep convolutional neural network architecture to classify culture disease has been proposed [20], along with a one-time detector that was used to identify and locate the sheet. A system based on the classifiers combination technique is proposed in a study [21]. This method included two variations of combinations [22]. Moreover, a methodology called IoT_FBFN was developed using Fuzzy Based Function Network (FBFN). This network had the computational power of fuzzy logic and the learning agility of the neural network to achieve greater precision in identifying and classifying gals relative to alternative approaches. An enhanced artificial neural network-based approach is proposed in another study [6]. The author extracted pixel values and feature values and used them as input to the enhanced ANN for image segmentation. After deve]loping a CNN-based model, segmented images were given as input into the proposed CNN model for image classification. An adaptive neural-fuzzy inference system and case-based reasoning for earlier identification of banana diseases are proposed in a study [23]. The modified K-mean segmentation algorithm is generated, which is used to separate the target area from the background in the rice plant image as proposed in a study [24]. An method was introduced to monitor natural habitats using Remote sensing, deep learning, and computer vision. The research team has leveraged remote sensing data of the last 20 years and deep-learning methods to enable precise image recognition. The computationally efficient approaches of CNN configuration are also demonstrated [25].

## DATA AND METHODOLOGY

The entire experiment is conducted in three steps, namely: Data Acquisition, Data pre-processing and Model building and validation. The flow diagram of the approach is shown in Fig. (1) and this section contains a detailed description of these steps.

### Data Acquisition

The plant diseases, Rust and Blast are the major diseases responsible for significantly reduced yield in pearl millet. These diseases were selected to build the model. The images of pearl millet rust and blast diseases were collected from open-source Kaggle [7]. The downloaded dataset consists of 4975 images belonging to two different classes. All the images used as input to build the model belong to the RGB color space and are in the JPG format. Below are the sample images for both Blast Fig. (**2**) and Rust Fig. (**3**) Diseases.

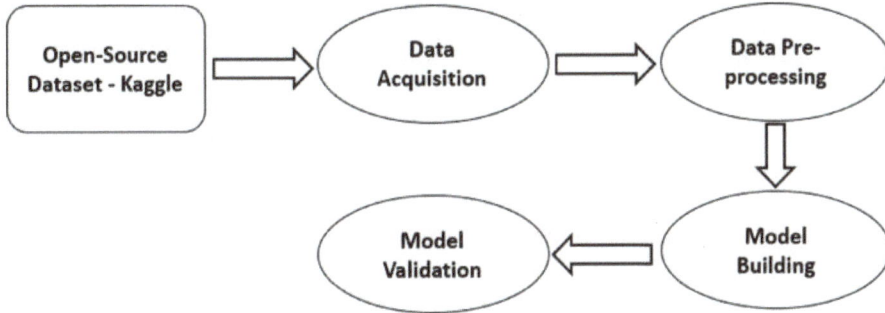

**Fig. (1).** Analytical life cycle.

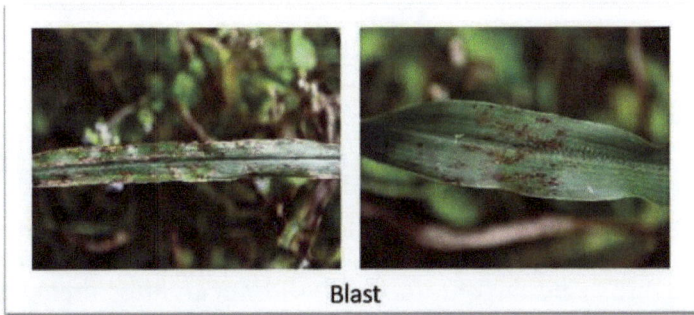

**Fig. (2).** Blast disease example.

**Fig. (3).** Rust disease example.

## Data Pre-processing

The downloaded dataset consisted of photographs with less noise and therefore, noise removal will no longer become an essential preprocessing step. Here we followed, only resizing of images as the essential preprocessing step. Resizing is a method in which all pieces of the matrix seem like distinct insulated pixels in the displayed image. In the real color image, the first level characterizes the depth of red pixels, the second plane denotes the depth of green pixels, and the last level symbolizes the depth of a blue pixel. To alter the dimensions of an image, we use the above approach to demonstrate two behaviors in the column. If the input image has extra dimensions, only the primary two dimensions can be corrected [6]. Here, we rescaled every image in to 224 x 224 ratio scale, as follows (Equation **1**).

$$r(Img, scale) = \prod_{k=1}^{n}( Img_K )  \tag{1}$$

**Model Building and Validation**

The classification model was trained and built to differentiate between blast and rust diseases in pearl millet crops. CNNs are used to construct the classification model. The CNN's architecture consists of many constructing blocks, inclusive of convolution layers, clustering layers, and fully connected dense layers. CNN is based on three foremost modules: convolution layers, pooling layers, and activation functions. A typical architecture is the repetition of a heap of multiple convolution and pooling layers, trailed by a fully connected dense layer or layers. Every block contains a convolutional layer, an activation layer and a maximum pooling layer. Three of those blocks observed *via* fully connected layers and sigmoid activation are used in our model building architecture. The feature vectors are obtained using Convolutional and pooling layers, whereas fully interconnected layers are consumed for class categorization. The non-linearity in the model was introduced through Activation functions in every layer.

A convolution layer is the most essential building block of the CNN architecture that performs feature extraction, which usually includes linear and nonlinear actions, *i.e.*, convolution action and activation function. With the increase in depth, the complexity of the extracted features also increases. The architecture of CNNs used to build models is depicted in Fig. (**4**). The scale of the filter is set to 2 x 2 whereas numbers of filters are accelerated gradually as we travel from one block to the next block. Initially, 16 filters are used in the first convolutional block, which is progressively raised to 32 in the second, and then to 64 in the final two blocks. This raise in the number of filters is essential to catch up for the discount on the size of the feature vectors produced *via* pooling layers in every block. We used the max pooling layer to decrease the number of feature vectors, speed up the training, and to make sure the model is less susceptible to small variations in input. The kernel size for max pooling was fixed at 2x2. The nonlinearity in every block was introduced by the ReLU activation layer (Equation **2**).

$$\text{Output} = \text{Max (Zero, Input)} \tag{2}$$

Additionally, the Dropout regularization approach is used with retains probability of 0.5 to prevent the model from over-fitting on train data. Drop-out regularization randomly ignores some nodes during model building at every iteration to decrease the variance of the model and thereby simplify the network. It also helps in the prevention of over-fitting. Finally, the output block comprises both categories of fully connected layers. The dense layer is aided by sigmoid activation Fig. (**5**) function to calculate the probability values for both classes.

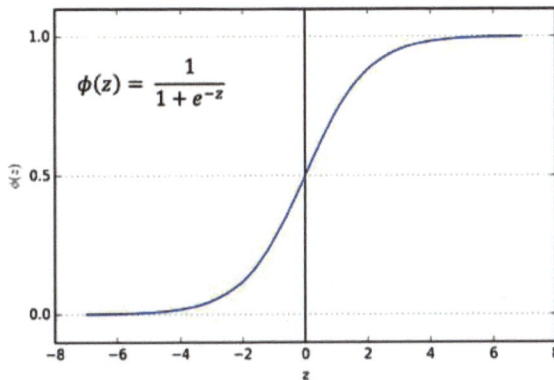

$$\phi(z) = \frac{1}{1+e^{-z}}$$

**Fig. (4).** Sigmoid Activation Function.

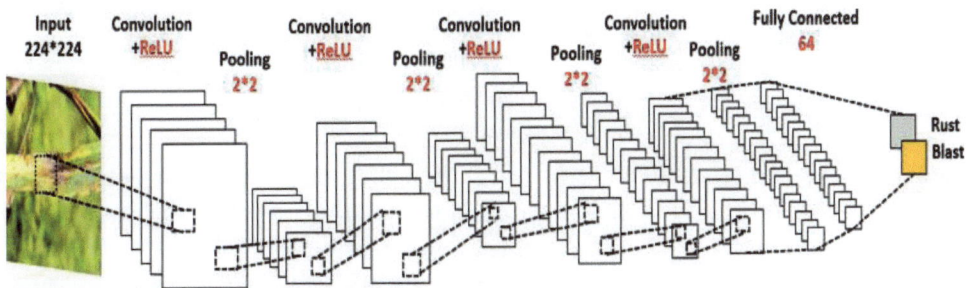

**Fig. (5).** Convolutional Neural Network Architecture.

The above-proposed technique has been applied to classify the pearl millet diseases blast and rust. A neural network library, Keras, authored in Python, is used to train and implement the model. Total 4975 images were used to train and validate the model. Of these, 416 images were kept aside for validating the model, and all the remaining images (4559) were used to train the model. To enhance the dataset, data is automatically augmented with the help of augmentation methods by randomly rotating the images slightly, horizontal reversing, and vertical and horizontal flipping of images. Binary cross entropy is used as a loss function with RMSprop as an optimizer to optimize the learnable parameters. The model was trained for 20 epochs with a batch size of 50 images.

## Evaluation Metrics

The metrics used for evaluating the classifier models are derived from the confusion matrix and the metrics are Accuracy, Precision, Recall and F1 Score. A detailed explanation of these metrics is given below.

**Confusion Matrix:** It is a table often used to measure the performance of machine learning classification problems, wherever output can be of two or more classes. It measures the number of rightly and wrongly classified samples into respective ground truth classes. In our study, True Blast represents rightly classified images, while False Blast denotes wrongly classified images of blast disease. Similarly, True Rust denotes correctly classified images, and False Rust represents incorrectly classified images of rust disease. The confusion matrix for our disease classification is represented in Table **1**. Based on the labels presented in Table **1**, we defined the evaluation metrics.

**Table 1. Pearl Millet Disease Confusion Matrix.**

| Actual | | Predicted | |
|---|---|---|---|
| | | Blast | Rust |
| | Blast | True Blast | False Rust |
| | Rust | False Blast | True Rust |

**Accuracy (A):** It is the measure of how often the classifier is correct, the higher the value the better the model. The formula to calculate is given in (Equation **3**).

$$A = \frac{True\ Blast + True\ Rust}{Sum\ total\ of\ all\ Blast\ and\ Rust} \tag{3}$$

**Precision (P):** It measures from all the classes we have predicted as blast; how many are actually a blast. It is calculated using (Equation **4**).

$$P = \frac{True\ Blast}{Sum\ total\ of\ Predicted\ as\ Blast} \tag{4}$$

**Recall (R)**: This is the measure of correct identification of samples of the blaster class from the total number of samples of that class. It measures all the Blast; how many are predicted correctly as Blast (True Blast). It is calculated using the formula given in (Equation **5**).

$$R = \frac{True\ Blast}{Sum\ total\ of\ Blast} \tag{5}$$

**F1 Score (F1)**: This measures test accuracy. The Harmonic Mean (HM) between Precision and Recall is called the F1 Score. The formula to calculate F1 Score is given in (Equation **6**).

$$F1 = HM\ (P,\ R) \tag{6}$$

## RESULTS AND ANALYSIS

The deep learning frameworks Keras/Tensorflow are used to train the model, on a laptop Intel Xeon E3-1505M v5. The model performance was evaluated based on the accuracy metrics. At 20 epochs of training, we achieved the highest validation accuracy of 98.08%, where a 97.59% of training accuracy was reported. The different evaluation metrics are presented in Table **2**.

**Table 2. Evaluation Metrics.**

| Matrix | Train | Validation |
| --- | --- | --- |
| Recall | 97.48% | 98.39% |
| Precision | 96.40% | 97.34% |
| Accuracy | 97.59% | 98.08% |
| F1 Score | 96.93% | 97.86% |

Epochs wise the graphical representation of training and validation accuracy and loss are shown in Fig. (**6**). From the graph it is inferred that the model becomes stable after 12 epochs. From Table **2** and Fig. (**6**), it arrives at the conclusion that the model has performed well on both the training and validation datasets and can be used to classify pearl millet rust and blast diseases.

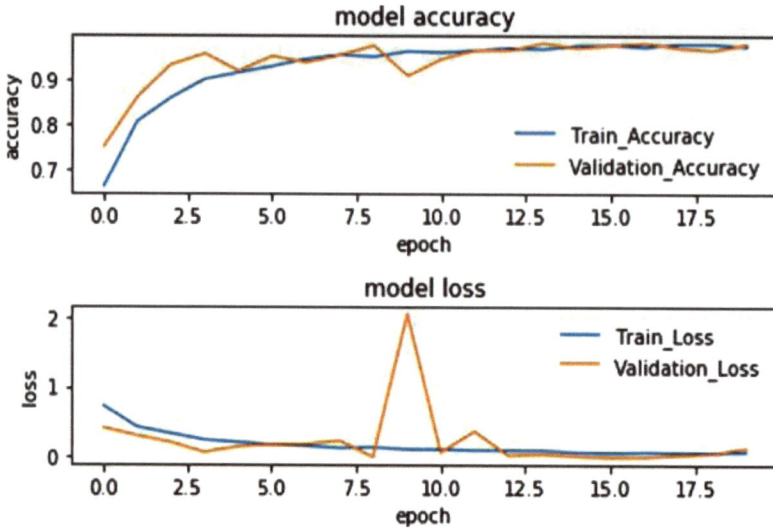

**Fig. (6).** Model accuracy and loss.

## CONCLUSION AND DISCUSSION

Agriculture is the most important sector on which most of the world's people depend for both their livelihoods and their food. Pearl millet is one of the primary food crops in arid and semi-arid zones. The document aims to classify and identify rust and murmur diseases of pearly millet. The methodology proposed above uses the CNN algorithm to differentiate rust and shiny leaf diseases from the pearl millet harvest from the data collected at Kaggle. The simple CNN architecture as represented in Fig. (5) is used to classify the pearl millet leaf diseases into blast and rust categories.

To improve the model in the future, several hyperparameters can be tuned for experimenting with the proposed model. In addition, advanced CNN algorithms may be used to enhance this model. Newly developed architectures can also be used to experiment with improving the overall accuracy of the model. The same algorithm could also be extended to build new models to identify and classify the other diseases of pearl millet. The above-built model is useful as a tool to help farmers in detecting rust and blast diseases of pearl millet. The proposed model accurately detects pearl millet diseases with an accuracy of 98% and with very less computational power.

## ACKNOWLEDGEMENTS

All individuals listed as authors have contributed substantially to the design, performance, analysis, or reporting of the work.

# REFERENCES

[1]     S.A. Miller, F.D. Beed, and C.L. Harmon, "Plant disease diagnostic capabilities and networks", *Annu. Rev. Phytopathol.,* vol. 47, no. 1, pp. 15-38, 2009.
[http://dx.doi.org/10.1146/annurev-phyto-080508-081743] [PMID: 19385729]

[2]     N. Ahmad, H.M.S. Asif, G. Saleem, M.U. Younus, S. Anwar, and M.R. Anjum, "Leaf Image-Based Plant Disease Identification Using Color and Texture Features", *Wirel. Pers. Commun.,* vol. 121, no. 2, pp. 1139-1168, 2021.
[http://dx.doi.org/10.1007/s11277-021-09054-2]

[3]     R. Alguliyev, Y. Imamverdiyev, L. Sukhostat, and R. Bayramov, "Plant disease detection based on a deep model", *Soft Comput.,* vol. 25, no. 21, pp. 13229-13242, 2021.
[http://dx.doi.org/10.1007/s00500-021-06176-4]

[4]     J. Chen, J. Chen, D. Zhang, Y.A. Nanehkaran, and Y. Sun, "A cognitive vision method for the detection of plant disease images", *Mach. Vis. Appl.,* vol. 32, no. 1, p. 31, 2021.
[http://dx.doi.org/10.1007/s00138-020-01150-w]

[5]     S.S. Chouhan, U.P. Singh, and S. Jain, "Applications of Computer Vision in Plant Pathology: A Survey", *Arch. Comput. Methods Eng.,* vol. 27, no. 2, pp. 611-632, 2020.
[http://dx.doi.org/10.1007/s11831-019-09324-0]

[6]     T.M. Prajwala, "Tomato leaf disease detection using convolutional neural networks", *Proceedings of 2018 Eleventh International Conference on Contemporary Computing (IC3),* 2018 Noida, India

[7]     N. Kundu, G. Rani, V.S. Dhaka, K. Gupta, S.C. Nayak, S. Verma, M.F. Ijaz, and M. Wozniak, "IoT and interpretable machine learning based framework for disease prediction in pearl millet", *Sensors,* vol. 21, no. 16, p. 5386, 2021.
[http://dx.doi.org/10.3390/s21165386]

[8]     J. Boulent, S. Foucher, J. Théau, and P.L. St-Charles, "Convolutional neural networks for the automatic identification of plant diseases", *Front. Plant Sci.,* vol. 10, no. 941, p. 941, 2019.
[http://dx.doi.org/10.3389/fpls.2019.00941] [PMID: 31396250]

[9]     S. Vallabhajosyula, V. Sistla, and V.K.K. Kolli, "Transfer learning-based deep ensemble neural network for plant leaf disease detection", *J. Plant Dis. Prot.,* no. 426, 2021.
[http://dx.doi.org/10.1007/s41348-021-00465-8]

[10]    V.K. Vishnoi, K. Kumar, and B. Kumar, "A comprehensive study of feature extraction techniques for plant leaf disease detection", *Multimedia Tools Appl.,* vol. 81, no. 1, pp. 367-419, 2022.
[http://dx.doi.org/10.1007/s11042-021-11375-0]

[11]    V.K. Vishnoi, K. Kumar, and B. Kumar, "Plant disease detection using computational intelligence and image processing", *J. Plant Dis. Prot.,* vol. 128, no. 1, pp. 19-53, 2021.
[http://dx.doi.org/10.1007/s41348-020-00368-0]

[12]    S.F. Syed-Ab-Rahman, M.H. Hesamian, and M. Prasad, "Citrus disease detection and classification using end-to-end anchor-based deep learning model", *Appl. Intell.,* pp. 927-938, 2021.

[13]    R.K. Singh, A. Tiwari, and R.K. Gupta, "Deep transfer modeling for classification of maize plant leaf disease", *Multimedia Tools Appl.,* vol. 81, no. 5, pp. 6051-6067, 2022.
[http://dx.doi.org/10.1007/s11042-021-11763-6]

[14]    M. Shantkumari, and S.V. Uma, "Grape leaf segmentation for disease identification through adaptive Snake algorithm model", *Multimedia Tools Appl.,* vol. 80, no. 6, pp. 8861-8879, 2021.
[http://dx.doi.org/10.1007/s11042-020-09853-y]

[15]    N. Sengar, M.K. Dutta, and C.M. Travieso, "Computer vision based technique for identification and quantification of powdery mildew disease in cherry leaves", *Computing,* vol. 100, no. 11, pp. 1189-1201, 2018.
[http://dx.doi.org/10.1007/s00607-018-0638-1]

[16] C. Pandey, N. Baghel, M.K. Dutta, A. Srivastava, and N. Choudhary, "Machine learning approach for automatic diagnosis of Chlorosis in Vigna mungo leaves", *Multimedia Tools Appl.,* vol. 80, no. 9, pp. 13407-13427, 2021.
[http://dx.doi.org/10.1007/s11042-020-10309-6]

[17] S. Nandhini, and K. Ashokkumar, "An automatic plant leaf disease identification using DenseNet-121 architecture with a mutation-based henry gas solubility optimization algorithm", *Neural Comput. Appl.,* vol. 34, no. 7, pp. 5513-5534, 2022.
[http://dx.doi.org/10.1007/s00521-021-06714-z]

[18] V. Muthukrishnan, S. Ramasamy, and N. Damodaran, "Disease recognition in philodendron leaf using image processing technique", *Environ. Sci. Pollut. Res. Int.,* vol. 28, no. 47, pp. 67321-67330, 2021.
[http://dx.doi.org/10.1007/s11356-021-15336-w] [PMID: 34245416]

[19] G. Mukherjee, B. Tudu, and A. Chatterjee, "A convolutional neural network-driven computer vision system toward identification of species and maturity stage of medicinal leaves: Case studies with Neem, Tulsi and Kalmegh leaves", *Soft Comput.,* vol. 25, no. 22, pp. 14119-14138, 2021.
[http://dx.doi.org/10.1007/s00500-021-06139-9]

[20] R. Gajjar, N. Gajjar, V.J. Thakor, N.P. Patel, and S. Ruparelia, "Real-time detection and identification of plant leaf diseases using convolutional neural networks on an embedded platform", *Vis. Comput.,* 2021.
[http://dx.doi.org/10.1007/s00371-021-02164-9]

[21] I. El Massi, Y. Es-saady, M. El Yassa, and D. Mammass, "Combination of multiple classifiers for automatic recognition of diseases and damages on plant leaves", *Signal Image Video Process.,* vol. 15, no. 4, pp. 789-796, 2021.
[http://dx.doi.org/10.1007/s11760-020-01797-y]

[22] S.S. Chouhan, U.P. Singh, and S. Jain, "Automated plant leaf disease detection and classification using fuzzy based function network", *Wirel. Pers. Commun.,* vol. 121, no. 3, pp. 1757-1779, 2021.
[http://dx.doi.org/10.1007/s11277-021-08734-3]

[23] A. Athiraja, and P. Vijayakumar, "RETRACTED ARTICLE: Banana disease diagnosis using computer vision and machine learning methods", *J. Ambient Intell. Humaniz. Comput.,* vol. 12, no. 6, pp. 6537-6556, 2021.
[http://dx.doi.org/10.1007/s12652-020-02273-8]

[24] K.S. Archana, S. Srinivasan, S.P. Bharathi, R. Balamurugan, T.N. Prabakar, and A.S.F. Britto, "A novel method to improve computational and classification performance of rice plant disease", *J. Supercomput.,* no. 0123456789, 2021.
[http://dx.doi.org/10.1007/s11227-021-04245-x]

[25] D. Dwivedi, and G. Patil, "Lightweight convolutional neural network for land use image classification", *Journal of Advanced Geospatial Science & Technology,* vol. 1, no. 1, pp. 31-48, 2022.https://jagst.utm.my/index.php/jagst/article/view/31

## CHAPTER 4

# Artificial Intelligence-based Solar Powered Robot to Identify Weed and Damage in Vegetables

**Kitty Tripathi**[1,*] and **Sushant Bhatt**[2]

[1] *Department of Electrical Engineering, Babu Banarasi Das Northern India Institute of Technology, Lucknow, India*

[2] *Shri Ramswaroop Memorial University, Lucknow, India*

**Abstract:** The agriculture sector plays a vital role in the Indian Economy and is known as one of the key areas where automation is emerging to enable farmers to increase the yield, prevent damage to the crop, reduce harvesting cost, *etc*. Artificial Intelligence (AI) offers a large number of direct applications across various sectors and it can bring a paradigm shift in the Indian farming sector. According to the report of the United Nations, the land area for cultivation will be 4% by the year 2050 so smart farming processes are the need of the hour and AI can help in finding solutions to increase the yield of crops and ensure food security. The chapter focuses on the role of solar-powered robots in the agriculture sector with the application of computer vision which is capable of recognizing the physical properties of vegetables and helps in monitoring the yield. We analyse a vegetable image data set with mass and dimension values collected using a computer vision system and accurate measuring devices. After successful detection and instance-wise segmentation, we extract the real-world dimensions of the selected vegetable. After monitoring the health of vegetables, the robot shares the profile through IoT in real-time and thus with low labour cost and without exhaustive search, the crop can be prevented from damage by weeds which can be identified at an early stage. Initial evaluation of the developed prototype exhibited a noteworthy potential of this system in the area of effective control of weeds and crop damage and assisting in harvesting.

**Keywords:** Plant disease detection, Pearl millet leaves, Deep learning, Machine learning, Convolutional neural network.

## INTRODUCTION

India is known to be an agricultural country due to the fact that agriculture and its allied industries still act as one of the main resources of income for the majority of

* **Corresponding author Kitty Tripathi:** School of Computer Science & Engineering, Galgotias University, Buddha International Circuit, Greater Noida, Uttar Pradesh, India; E-mail: sersk2006@gmail.com

**Praveen Kumar Shukla & Tushar Kanti Bera (Eds.)**

the population in rural India. It also is an essential sector that affects the Indian economy and adds around 17% of the total GDP of the nation.

India is the largest Global manufacturer of milk, pulses and jute, and is the second-largest manufacturer of rice, wheat, sugarcane, groundnut, vegetables,

fruit, and cotton. It is furthermore one of the most important producers of spices, fish, poultry and livestock, and plantation crops [1].

For agriculture, the environment in India has huge variations that differ from extreme humid conditions to tropical regions that are very dry in the southern region of the country. The northern region on the other hand has temperate and high-altitude which results in a great diversity of ecosystems. Despite the fact that the agricultural sector is accomplishing abundance in food when it comes to production, still India plays a significant role in providing food for a part of the world's hunger and thus have a responsibility to provide over 190 million undernourished people [2].

Since Indian Agriculture is a resource of the various inventive ways of irrigation, it now raises serious issues of sustainability and thus it is very essential that there is improvement in its management of agricultural practices on numerous facades. With the rapid spread of the Digital India movement, the Indian Agri-sector is also rapidly changing and is readily accepting automation in the sector and Artificial Intelligence is the main technology that is moving from the conventional methodologies and traits within the latest years [3].

AI is a data-driven technology and a large amount of ever-changing data lead to uncertainty, inaccuracy, and contradictions. Currently, the systems for approaching features are typical with the assistance of utilization of iterative techniques, and the neural community architectures related to each difference, which are together studied under the terms "Deep Learning" and "Machine Learning" [4, 5].

The application of robotics in agriculture might be an innovative step to change the productivity of labour. By either mimicking human skills or escalating them, robots can help in removing the critical human constraints which include the ability to operate in difficult agricultural environments [6, 7].

The quality of vegetables depends on the aspect of the vegetable image which is of morphological features. Colour is the primarily used feature to distinguish whether the vegetable is affected by weeds or not. Out of the various factors like colour, size, and texture and shape, colour is the most important feature which indicates a high impact on the quality of the vegetable.

## DIGITAL AGRICULTURE: IMPACT & CHALLENGES

In spite of the fact that a large segment of the population is involved in agriculture in India yet, there are a lot of challenges. In recent years, there is an increase in the shortage of labour and youth participation in the sector [8]. Indian agriculture also faces inadequate resources and information on mechanization. So there is an increased need for good equipment and improved techniques for increasing production in terms of both quality and quantity.

The practices in agriculture are majorly dependent on the factors like climatic conditions, water level, temperature, fertility of the soil, forecasting of weather, availability of fertilizers, use of pesticides, presence of weeds, and the method of harvesting.

In many of the cases, the farmers in India use the conventional methods to predict crop yield which is dependent on the knowledge that is based on their previous experiences, nonetheless, this approach alone possibly will not be an efficient way of prediction because climatic conditions have drastic changes regularly due to the overall change in the weather forecast at the global level. For addressing this issue, there can be an additional scientific practice with the advent of technology which is known as agro-based big data analytics [9]. Big data analysis can be used as an opportunity which can be used to analyse crucial factors which will help in controlling the crop yield. This method can also be used for the analysis of social, economic, and political impacts on the success rates of the different agricultural practices.

For better yield of crops, one of the basic solutions is to increase the overall cultivable land that is appropriate for the growth of a particular crop. It also requires a reduction in the damage of the crop and a decrease in the net operating cost by the implementation of various upright agricultural practices.

To improve the crop yield, there is a need of controlling the key factors that are used in the Indian agricultural practice including the type and quantity of fertilizer, the level of water and the resource from where it is obtained, seed quality that is used in cropping, reduction of the biotic stress which is caused by weeds and pests.

There is also a practice of manual inspection of the field by the farmers and the removal of weeds and contamination found if any is done manually which is a less effective approach and it does not produce any significant limitation in support of higher crop yield so it can be replaced by sensor network mounted in the field which can effectively understand the needs of the crop at real-time [10 - 12].

Although there is a lot of mechanization of agriculture in some parts of the country, yet most of the operations in the agriculture sector are carried out manually with conventional tools and implementation methodology. There is very low or no use of machines in the processes like plugging, sowing, irrigating, thinning and pruning, weeding, harvesting threshing, and transporting the crops. This thus becomes intensive and requires a lot of labour which does not even result in a good yield per capita of the labour force.

The digital farming system will enable the farmers to get informed and be knowledgeable about the various methods of cultivation which are used in various parts of the world for a particular kind of crop or vegetable so that they can be equipped with advancements in technologies in the sector. In such a manner, the farmers can use their inheriting practices of farming along with useful information and equipment to increase productivity.

They can also get aware of their historical data which can help in understanding the various conditions and difficulties so that there is essential information available that will help them in making the right decision.

Various kinds of techniques are available where the historical data of any farm can help in the prediction of conditions like the existence of weeds, and pests and can thus help in balancing the necessities of organizing a profitable business by considering the following points:

   i. Soil management & crop nutrition.
   ii. Crop rotation.
   iii. Energy consumption.
   iv. Waste and pollution management system.
   v. Organization management.
   vi. Agriculture site.
   vii. Monitoring & auditing.
   viii. Crop protection.
   ix. Wildlife and landscape management.

**INTRODUCTION TO ROBOTICS**

Robotics has its own significance in the area of science and fiction. Initially, the word Robotics is found in a play written in Czechoslovakia around 1920 but it took around 40 years to become a part of the current high-tech technology of industrial robotics and automation. Nowadays, robots are controlled by computers [13].

## Robotics

Robotics is applied engineering which includes computer programming, automation, elements of machine design, mechatronics, artificial intelligence, and a combination of machine tool technology. The design of Robots is very essential and useful in daily life applications in various industrial operations where automation manufacturing systems are used [12, 13].

The main objectives of a robotic system are:

• To increase the productivity of the system.

• Reduce production life.

• Robotics minimize the manpower requirement.

• It enhances the quality of products.

• Minimize loss of work time.

• Highly reliable and provides a high-speed production rate.

There are various applications of robots:

• In the Material Handling System

• In loading and unloading of goods

• In various LASER operations,

  a) In Assembly tasks, assembly cell designs, parts mating.

  b) During Industrial Inspection.

• Used in various processing operations like,

  a) Welding

  b) Spray painting

  c) Spray Coating

- In machining operations

- In the routing process.

- In the grinding process.

- In polishing debarring wire brushing.

## Need of Robotics

Industrial robots play an important role in the automated manufacturing system.

There are various needs for robotics:

• Robotics can provide more speed and accuracy during manufacturing.

• Repetitive tasks can be done with better quality and consistency by robotics.

• No negative attributes of human work like fatigue, need for rest, and diversion of attention.

• Very accurate and improved working conditions.

• Robotics reduces the chance of an industrial accident.

## Industrial Robots

The official definition of an industrial robot is provided by the Robotics Industries Association (RIA). An Industrial robot is defined as an automatic, freely programmed, servo-controlled, multi-purpose manipulator to handle various operations of an industry with variable programmed motions.

## Automation and Robotics

Automation and robotics are two technologies that are much related to each other. Automation is a technology that is related to the use of mechatronics and computer-based programming. Automation is used to operate and control production in industries.

In industries, assembly machines, Numerical control-based machines, CNC, and transfer lines are operated by automation [14]. In other words, Robotics is a form of industrial automation.

The specifications of the robots are as under,

• Axis of motion for robots

• Workstations for robots

• Speed of robots

• Acceleration of robots

• Payload capacity of robots

• Accuracy of robots

• Repeatability of robots

The future of robots is based on the following activities:

• In intelligence-related works

• Medical care and hospital duties *etc.*

• Household robots

• In fire fighting operations

• In underground coal mining

• In hazardous environments

• In sensor capabilities

• In Flexible Manufacturing Systems

• In tele-presence

• In the mechanical design of various manufacturing components

• In mobility and navigation

• In networking and multi-level marketing

• Robots in space

• In the security of various buildings and sites

• Garbage collection and waste disposal operations

## Control Systems for Robotics

In robotics, the motion control system is used to control the movement of the end-effector or tool [13]. On the basis of controlling, robots can be divided into the following types:

• Limited sequence robots (Non-servo)

• Playback robots with point to point (servo)

• Playback robots with continuous path control,

• Intelligent robots.

### *Limited Sequence Robots (Non-Servo)*

These robots do not provide the servo-controlled operation to inclined relative positions of the joints. They are regulated by limit switches and mechanical stops. These types of robots are generally used in simple motion as pick and place operations.

### *Point to Point Motion*

The velocity acceleration and path of motions of Robots from start to end of the path are controlled by these types of robots. PLCs (Programmable Logic Controllers) are used to control these types of motion. They perform the motion cycle that consists of a series of desired point locations.

### *Continuous Path Motion*

Continuous path motion robots are capable of performing motion cycles in a controlled path. These types of robots move through a described and desired path.

### *Intelligent Robots*

These robots are programmable and they are also interacting with their environment and giving their intelligence towards work. They make logical decisions based on sensor data received from the operation.

### Presence of Movement for Robots in the Agriculture Sector

The movement of Robots is another measure of performance which is defined in the below given three features:

• Spatial resolution

• Accuracy

• Repeatability

## AN INTRODUCTION TO SOLAR ENERGY

The sources of traditional and non-renewable energy such as coal and petroleum products are consumed by human beings very rapidly. The formation of these conventional fuels takes a very long process to recreate. Hence there is a need for an alternative source of energy such as Wind Energy, Bio-energy, and Solar Energy. These are non-conventional or renewable energy sources. Amongst the renewable energy sources solar energy is considered to be the best and most popular source of energy which is abundant in nature and the availability of solar energy is free of cost [15]. Solar energy is produced by the use of solar cells. Using the principle of the photovoltaic effect, these solar cells convert sunlight into electrical energy. The electricity can directly be used to charge the batteries used for various appliances.

The advantages of solar panels with robotic systems are as under,

• No external fuels are required.

• Sunlight is available free of cost.

• Solar modules have a very long life.

• Solar panels must be operated at very high temperatures.

• This is free from Pollution.

• Minimum Maintenance.

• Independent working

• Noise-free operation as there are no moving parts.

• Efficient in remote areas and hilly places.

• High power generation capacity.

• It can be installed and mounted easily with minimum cost.

The disadvantages of solar panels in robotics systems are as under,

• Very high initial cost.

• The whole operation of solar energy depends upon sunlight.

• Battery storage cost is additional.

• The output of the solar panel depends upon the climatic condition, location, latitude, longitude, altitude, tilt angle, aging, dent, *etc*.

• To install solar panels, a large area is required.

## Photovoltaic Effect on Solar Generation

In the photovoltaic effect, electricity is generated with the help of direct sunlight. The absorption of ionizing radiation produces an electromotive force. The best performance of the photovoltaic effect is seen in silicon.

### *Solar Cell: Construction and Working*

The Solar Cell is a semiconductor device. It consists of a simple p-n junction that produces DC electricity when exposed to sunlight directly. The "Semi-Conductor" materials are used in the formation of Solar cells. Single crystal silicon, polycrystalline and amorphous Silicon are used in the manufacturing of solar cells. The P-N junction of the solar cell gives rise to diode characteristics. A solar cell is a PN junction device on which front and back electrical contacts are screen-printed.

The solar power system consists of the following components:

• Solar array.

• Battery Bank

• Solar Charge Controller

• Field Junction Box

• Solar Module Mounting Structure

• Earthing kit

• Cables.

There are various types of solar panels used in this process. The Crystalline Silicon panels may be in the form of Mono, Poly, and Amorphous types. There are different sizes or Areas of cells available in panels. They may be rectangular/Circular/Square/Pseudo-square/Semi-circular *etc.* The power of the panel may be in high, medium, and low ranges.

## LOAD CALCULATION OF SOLAR PANELS

### For DC Loads

(Load Amps)*(Operating Hours per day) = Amp Hour per Day (AHPD)

### For AC Loads

(AC Watts)*(Operating Hours per Day) = Watt Hours per Day (WHPD)

(AHPD = WHPD)/(Inverter Efficiency)*(Nominal System Voltage)

**Sizing Solar Array** No. of Series Modules = Nominal DC System Voltage / Nominal Voltage of Solar Module

No. of Parallel Modules = AHPD/ [(1 – module derating*) *(Battery Efficiency) *($I_m$ of module)* (Solar insolation of worst month)] No of Modules in Solar Array = (No of Series Modules) *(No of Parallel Modules)

Capacity of Solar Array = (No of Modules)*(Capacity of Each Module)

### Deciding Battery capacity

Battery Capacity (in AH) = (AHPD * No. of back up days) / Max DOD

No of series Batteries = System DC Voltage / Battery Voltage

No of Parallel Batteries = Total AH Required / AH of Individual Battery

## SAMPLE SYSTEM DESIGN

**Step 1:** There is the need to determine the DC load by the formula given below:

DC load of a device 1 = (No. of DC devices)*(Device Watts)*(Hours of daily use)

$$= DC\ Watt\ Hours\ per\ Day \qquad (1)$$

**Step 2:** When there is a need for determination of the AC load, we use the formula

AC load of a device 1 = (No. of AC devices)* (Device Watts)*(Hours of daily use(

$$= \text{AC Watt Hours per Day} \qquad (2)$$

**Step 3:** For obtaining the Total System load, we use the following formula with an example:

Suppose total DC load [A] = 4000 Watt Hours per Day

& total DC load [B] = 2000 Watt Hours per Day

Total System Load [A+B] =6000 Watt Hours per Day

**Step 4:** Determine Total DC Ampere Hours per Day *i.e.* Total System Load/System Nominal Voltage

**Step 5:** Thereafter we determine Total Ampere Hours per Day with Batteries by performing the calculation of Total Ampere Hours per Day(Battery efficiency X Module derating).

**Step 6:** Determine the Total PV Array Current by calculating the Total Daily Ampere Hour requirement / Design Insolation

**Step 7:** Select PV Module type.

**Step 8:** After the module is selected we then determine the Number of Modules in parallel using the formula -Total PV Array Current / (Module Operating Current)

**Step 9:** Determination of the Number of Modules in the Series is performed by System Nominal Voltage / Module Nominal Voltage

**Step 10:** Determine the Total Number of Modules by performing the Number of modules in parallel X Number of modules in the Series.

**Step 11:** Choose a Battery that is appropriate for the system. In general, there is the use of Flooded Lead Acid Battery Cells.

**Step 12:** Choose Charge Controller by performing the following calculation:

(No. of Parallel Modules) * Isc * 1.25 = Charge Controller Capacity

**Step 13:** Lastly there is a need for the selection of Inverter done by the following method:

AC Watts on Inverter = (Total AC Watts of AC devices)* 25% extra.

## AGRICULTURAL ROBOT

The deployment of various individual physical sensors in the field is tedious, costly and the maintenance is quite a challenge within the field so an unmanned portable robot Fig. (**1**) can be used which can be equipped with the necessary sensors so that the measurement of the growth time course of the vegetable can be monitored effectively [15 - 17].

**Fig. (1).** Carrier trolley setup of Robot.

The framework is supported by wooden platforms which have a wirelessly operated trolley as shown in Fig. (**2**). Solar panels are mounted on top of the structure. The battery, inverter, and control panel are placed above the chassis of the trolley and below the solar panel.

**Fig. (2).** Solar tracking of Robotic module.

This type of solar power unit is used in the agricultural field for the spraying of insecticide and irrigation purposes. These units can also be taken back home to power household appliances by electric vehicle or tractor when not in use in the agricultural field.

The solar panels are oriented towards the sun by a solar tracker. Solar trackers are used to calculate the analog values that are coming from LDR (Light Dependent

Resistors). A dual-axis solar tracking system regulated by automatic and manual control is implemented in our prototype. The single-axis rotation system is used in the automatic control system which rotates the panel horizontally (left to right) while the dual-axis rotation system rotates the panel in up and down position as shown in Fig. (**2**).

This design uses five numbers of solar panels in a series connection to generate 3V and 0.25 Amp current. This voltage and current are increased to 13.5 V and 0.1 Amp with the help of a 12 V DC supply. Both AC and DC supply is used in our model and this may be switched on in particular option the control panel and the rear view of the module is shown in the figure below.

This system uses ESP32-CAM which has three GND pins and two power pins. The ESP32-CAM is powered with 5V and produces an output of 3.3V. The same output reading of 5V can also be obtained when it is connected to a 3.3V supply. The ESP module operates on a pre-programmed Wi-Fi network. We can see the outputs of live visuals directly from the camera by opening the URL in a browser.

## Mechanical Design of Agricultural Robot

To achieve the objective of segregating the unhealthy and healthy conditions of vegetables in the field, the design of the robot has the following enlisted features Fig. (**3**).

**Fig. (3).** Rear view of the robotic Vehicle.

The outer frame of the robot is a wooden structure with an approximate weight of 3 .5 Kg.

The panels are mounted on the top at a height of 80mm through a fold-and-unfold mechanism which is made using an ejecting plate and a frame which is made of plastic and supported on the wooden structure by a wooden slab.

The unfolding of solar plates is controlled by a set of push buttons which are operated manually and positioned on either side of folding segments vertically as shown in the Figure below.

The dual-axis tracking mechanism of the solar plate modules is controlled by a DC geared motor with an operating voltage of 5V. This motor is supported by a thin rod with springs attached to it for the adjustment of height as shown in Fig. (**2**). The figure shows the motion of the panel assembly in the horizontal and vertical direction which is controlled by the geared DC motor. This system also has Light-dependent resistors mounted along the length and width of the array of solar panels which will help the system to identify the position of maximum light intensity and align accordingly. This system increases the efficiency of power generation by about 30% more than the static solar panel system.

There are 5 solar panels mounted on this structure as shown in Fig. (**4**). The panels used are mono-crystalline with dimensions of 158 mm * 158 mm ± 0.25 mm. The diagonal length is 212mm ± 0.25 mm and the thickness of the panel is 190 μm ± 30 μm. The module has a positive electrode width of 0.7 mm with a 116 Root fine grid and blue silicon nitride anti-reflection coating whereas for the back electrode, the width is 1.8mm and it also has an antireflective film.

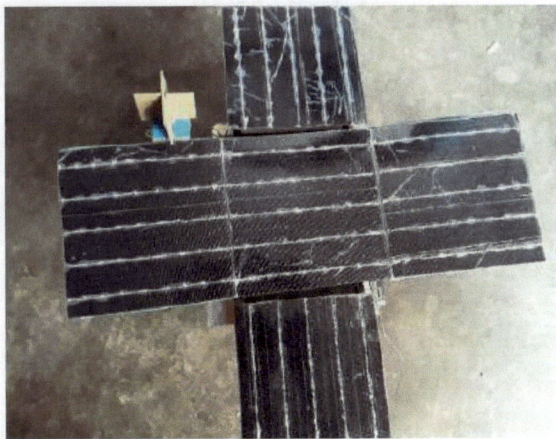

**Fig. (4).** Unfolded solar panels.

All the solar panels are connected in a series to increase the total generating voltage *i.e.* about 3.00V and generating current is about 250mA. Since it is not a sufficient value for charging a battery, hence a DC-DC step-up module (XL6009) is used which boosts the voltage to 13.5V DC.

The carrier trolley of the robot is a four-wheeled system with a DC-geared motor that is used to control the movement of the robot in all directions. The wheels used are made up of plastic and have a diameter of 70 mm and a width of 20 mm. The diameter of the shaft hole is 6 mm.

The main controller of this robot is Arduino Uno which consists of both a physical programmable circuit board and an Integrated Development Environment. It is a low cost widely used microcontroller with sufficient processing power for the application. The assembly has four pushbuttons, four LDR, and two toggle switches (T1 and T2) that are used for manual control and changing mode respectively. If the toggle T1 is LOW, manual mode is active such that the four pushbuttons change the direction of the solar panels. If T1 is active, automatic mode is in action and the program executes as follows:

• The microcontroller reads the analog values from four LDR sensors (top-right, top-left, bottom-right, bottom-left).

• In the next step, the microcontroller calculates the average values for top-bottom, left- right *i.e.* avgt_op, avg_left, avg_bottom, avg_right.

• Now, it will calculate the average values of difference; such as the difference between elevation and difference horizontal.

• If (difference horizontal) <= 10, the left-right motor turns the panel Left.

• If (difference horizontal)> 0, the left-right motor turns the panel right.

• Similarly, if (difference elevation) <= 10, top-down motor turns the panel Down.

• If (difference elevation)> 0, –top-down motor turns the panel's Top.

A camera module ESP32-CAM is used which has a development board that consists of an ESP32-S chip, an OV2640 camera, a micro SD card slot, and several GPIOs to connect peripherals.

For sharing the image data to the cloud, NodeMCU is used. An ESP8266 NodeMCU module is an open-source firmware and development board that is used for the application of the Internet of Things (IoT). It runs on Wi-Fi SoC from Espressif System. It can be powered through a USB port or a regulated supply of

3.3V. The microprocessor supports RTOS and operates at 80MHz to 160 MHz adjustable clock frequency. NodeMCU has 128 KB RAM and 4MB of Flash memory to store data and programs.

## WORKING OF SOLAR ROBOT

The main functioning module of the solar robot is Raspberry pi which is powered by a solar module. As solar power is unable to provide a continuous supply thus a rechargeable battery is used in this system to provide a consistent power supply to the vehicle. When any of the sensors become active, it sends an alert message to the cloud of NodeMCU. The Motor driver IC (L293D) then controls two DC motors concurrently in any direction. It works on the principle of the H Bridge and drives the motor in either clockwise or anticlockwise direction.

In the manual mode of operation, a web page is created for the control of the robotic vehicle. In case of any type of warning is sensed, it is viewed through the camera that is interfaced with Raspberry Pi so that the robot can be directed manually. Arrows indicating directional control can control the movement of the vehicle.

In automated mode information, various sensors (PIR Sensor, Temperature sensor, Ultrasonic sensor, Camera Module) are used as input. During the automatic mode of operation of the robot, GPS plays a major role in defining the current position and location of the robot for further navigation.

## COMPUTER VISION AND MACHINE LEARNING

Computer Vision provides a similar outcome as that of the human brain where vision is processed for sensing the surroundings. This technique is gradually expanding its application in the agriculture sector. This technique is not only helpful in increasing productivity but also helps in reducing production cost with the help of automation.

Many researchers have used the Concept of Machine vision in areas that include Harvesting, planting a crop or vegetable, checking the presence of weeds or unhealthy conditions, monitoring the farm, and analysis of weather conditions [18].

There has been a significant increase in the usage of unmanned aerial vehicles which are capable of flying autonomously. These are used majorly for getting information about the field by capturing images through cameras installed on it. These cameras are sometimes enabled with computer vision. When these machines are trained by machine learning, then they can be capable of detecting the various conditions of the field by using certain techniques for image segmentation and have an approach for detecting and identifying any and every object in the field.

The collected data from machine vision may also be used for the analysis of various critical factors of the field like the presence of nitrogen level and moisture in the field and various other parameters that can be used to train the machine so that prediction analysis can be made.

Machine learning-enabled computer vision is helpful in automating time consumption and monotonous tasks like sorting vegetables and fruits and identifying the longevity of the vegetable or fruit which will give an idea of where and how to send the fruit so that the quality remains intact.

They can not only identify the field conditions but can also be used for the automatic praying of fertilizers and pesticides by monitoring the infected crop in the field through computer vision.

Computer vision also provides efficient information in the phenotyping approach where image processing can help us obtain various useful information by colour enhancement, depth estimation, identifying segments of the field, *etc* [19].

In some of the cases, researchers have used spectroscopy technology for obtaining a spectral image from the data of computer vision to do the analysis and comparison of every spectral band and can use median filter pre-processing of the image which improves the accuracy of recognition of fruit [20].

## EVALUATING THE QUALITY OF VEGETABLES USING MACHINE VISION

The proposed design of the system is separated into five steps:

  i. Image for Input
 ii. Image pre-processing
iii. Segmentation of images
 iv. Extraction of Features
  v. Classification

Robotic vehicles are used for capturing images of vegetables in the image collection process. From these collected images, relevant information for unwanted growth for the estimation of weeds is collected and then used to carry out the collection of data. These data are now used for image pre-processing.

The image data is magnified by the image pre-processing method. This method removes the disinclined misrepresentations and the image features are expanded. This is required for setting up and processing a relevant image than the indigenous image. In image pre-processing, the chief features and assessment of vegetables can be completed by local pre-processing along with pixel pre-processing [18, 19].

When the pre-processing of images is completed, the process of image segmentation is in process. In image segmentation, the digital image is converted into well-defined areas. In the evaluation of fruit, image segmentation is used to separate the background for processing the important area. In further image analysis, the original segmentation is definite. The performance of the classifiers can be reduced with the help of improper segmentation. The different types of approaches include k-means clustering and fuzzy c-means which can be used for segmentation [20 - 22].

After the features are extracted, they can then be used for other analyses. The classification of vegetables along with the image clarification are considered in the extraction feature. The morphological and textural features are used for the analysis of the growth of weeds. The simulated assumptions of human prospects are helpful in determining the presence of damage or weed. The image of vegetables can establish a set of information or features like texture, colour, size and shape. These sets of information may be used in training sessions. After these training sets, the Neural Network algorithm is used to make a decision [23].

## CLASSIFICATION ALGORITHM

It is a part of supervised machine learning and is used where the output comes in the form of a category. Various kinds of classification algorithms can be used in the agriculture sector depending upon the requirement of information to be extracted from the image of the field.

The main issue is that the precision of the classifier depends on the prediction accuracy. In the dataset received from the computer vision, there is an option of splitting the data set into two-third amounts for training and the remaining for estimating the efficiency of prediction. Some of the researchers also divide the training sets into various subsets and for every individual subset, the classifier is trained individually and then the average error rate from each subset is obtained. It is one of the exhaustive computational methods but is very useful in determining the error rate of the classifier.

The agriculture data which is received from machine vision can be of both structured and unstructured data and hence the classification algorithm can easily perform the task of classification of the data into the respective classes. The classification model is used to determine the category or class of the data with the help of training in the system where the required features are extracted. The main steps of the classification algorithm are:

i. Initialization of the classifier is done.
ii. For a given training data "t" the trained label "l" is done by a fit method for the model.
iii. For any unlabelled data received, the predicted label is determined from the trained model.
iv. The model is then evaluated for error rate.

Logistic regression is one of the classification algorithms and is best suited for those agricultural data sets which have various independent variables that affect a single output variable. The main issue with this algorithm is that it is suited for those systems where the output prediction is done in binary variables. It does not give good accuracy when the data is having some missing values.

The Naïve Bayes algorithm is an algorithm that can be used for agricultural data that have features that are independent of each other. This algorithm does not require a large amount of training data for the estimation of the vital parameters in a shorter period. The only issue with this algorithm is that it does not have good accuracy when it works as an estimator.

One of the most convenient algorithms for the classification of agricultural data is Stochastic Gradient Descent which is very useful when the number of samples in the dataset is very large. The main issue with this algorithm is that it needs a large number of hyperparameters.

The K-Nearest neighbour algorithm is a very effective classification algorithm when it comes to datasets received from agriculture robots as this algorithm is quite robust to the data set that is very noisy. Since this algorithm classifies according to the majority vote at every point with the nearest neighbours, so it requires a large dataset for training.

Decision tree algorithms are used where the dataset is simple and does not need a lot of data preparation. This algorithm basically uses a sequence of rules which will contribute to the classification of data but it is quite an unstable method as a small disparity in the data may lead to a complex tree that cannot be a universal set for different cases.

Random forest classifiers are also used for classification where we need to reduce the over-fitting. This system uses a set of decision trees on various sample subsets of data with a sample size equal to the original sample to improve the accuracy of prediction. Its major drawback is that it cannot be used for real-time prediction. The Support vector machine is a method that can be used for agricultural data where the data is separated in space with wide gaps and separated into different categories. The testing data in this case is mapped in the same data space and then prediction is done which fits to a category.

## METHODS FOR COLOUR SELECTION AND EXTRACTION

The scattering of samples in the attribute space is regulated by sensor readings, as well as the colour channels for the distinct colour spaces in our system. The examination of the colour channels which become higher in relevance for the classification is essential to the analysers to discover the selection of boundaries among the two classes which are the health condition of vegetables and damaged condition. Features which are either consisting of noise or are irrelevant can increase the classification problems which may result in low performance even if the classification is accurate [24]. Selecting a subset of the k-most relevant features can be done manually or with the help of feature extraction tools.

It will aid in the elimination of redundant as well as irrelevant material. As a replacement for the original d-dimensional feature space, the generated k-dimensional feature space is used for categorization [25]. This is appropriate for assisting in the identification of a criterion for separating the two classes. Unlike feature extraction techniques, which try to construct new characteristic space, characteristic choice methods aim to select a subset of vectors that span a feature subspace that allows class. Unsupervised feature selection strategies are only dependent on the underlying distribution of samples, while supervised feature selection strategies incorporate class labels to discover the most relevant features [26]. The computation-intensive combinatorial challenge of selecting the subset of the most relevant attributes is a computationally expensive problem. The approach used for a rating or selecting features may influence how well an algorithm finds a good enough feature subset.

The purpose of feature selection and extraction can be determined by evaluating several points $X = x_1, x_2... x_N$, where xi Rd. The feature extraction and selection procedures generate a new set $X' = x'_1, x'_2,..., x'_k$, where $x'_i$ is the new dimension of the feature vectors, and k d seems to be the new element of the vectors [27].

The measurement of the $1^{st}$ order probability consists of evenness, variance, mean, standard deviation, RMS, inverse difference moment, kurtosis, and skewness which are statistical features. Here in this case, the histogram of the pixel intensity is utilized for determining the statistical features of the defective part. The features also comprise color features obtained from the mean and the standard deviation of RGB values.

The measurement of $2^{nd}$ order probability includes disparity, energy, entropy, consistency, and correlation which are textural features. The distribution of grey-scale in the grey image has an approximate reference of the texture of the image.

After extracting 108 features, the subset of 25 features is selected which include mean, standard deviation, homogeneity, entropy, variance, axis length, RMS, inverse difference moment, coarseness, homogeneity, correlation, centroid, orientation *etc*.

## CONCLUSION AND FUTURE SCOPE

In this work, solar operated robotic cart is proposed which can help in capturing the image of various states of vegetables growing in the field by a feed through a camera module mounted on the robotic vehicle. This vehicle is low cost and unmanned which will do the task of surveillance of the vegetable field to check the state of vegetables [31]. The image thus gathered is used to identify the damaged vegetable through the classification of image in two classes which are healthy condition and unhealthy or damaged condition of vegetable at the beginning of the development of the weed. The colour is the most important feature which was extracted by the image to identify and classify the state of the vegetable. The training was done by Artificial Neural network technique. Before characteristic segmentation and recognition, manual segmentation and sorting were undertaken. Till now the extraction of morphological features is done with different Machine Learning techniques like SVM, ANN, K-Means Clustering, KNN, and Fuzzy logic. Some of the researchers are now using deep learning algorithms and are using a fusion of colour and textured features of the image of a crop or vegetable or fruit for obtaining a higher degree of accuracy [28, 29]. The performance of classification is obtained through Precision, accuracy, sensitivity, specificity, FPR, and FNR. The maximum accuracy attained for k fold = 5/10/15 is 90.12%, 93.64%, and 94.48% respectively using SVM.

As a result, in the future, deep learning algorithms with several features combined for categorization could be considered to attain improved accuracy [30]. Statistics from fusion approaches such as classifier combination strategies, on the other hand, may improve the accuracy of forming decision demarcation between the two classes. The future scope of the solar power robotic vehicle is to classify live feed of images and if the damaged vegetable is observed or any weed growth is visualized, it can take the decision of cleaning the weed and sprinkling the fertilizers required through a bottle of fertilizer mounted on the unmanned vehicle as well.

## ACKNOWLEDGEMENTS

We are grateful to Babu Banarasi Das Northern India Institute of Technology and Shri Ram Swaroop Memorial University for all the support.

## REFERENCES

[1]　Food and Agriculture Organization, "India at a glance", Available From: https://www.fao.org/india/fao-in-india/india-at-a-glance/en/ (Accessed: January 3rd 2022).

[2]　World Internet Usage and Population Statistics, Available From: https://www.internetworldstats.com/stats.htm (Accessed: December 12th 2021).

[3]　B.A. King, T. Hammond, and J. Harrington, "Disruptive technology: Economic consequences of artificial intelligence and the robotics revolution", *Journal of Strategic Innovation and Sustainability,* vol. 12, no. 2, 2017.
[http://dx.doi.org/10.33423/jsis.v12i2.801]

[4]　J.M. McKinnon, and H.E. Lemmon, "Expert systems for agriculture", *Computers and Electronics in Agriculture,* vol. 1, no. 1, pp. 31-40, 1985.

[5]　J.L. Ruiz-Real, J. Uribe-Toril, J.A. Torres Arriaza, and J. Valenciano de Pablo, "A Look at the past, present and future research trends of artificial intelligence in agriculture", *J. Agron.,* vol. 10, no. 11, p. 1839, 2020.

[6]　K.G. Liakos, P. Busato, D. Moshou, S. Pearson, and D. Bochtis, "Machine learning in agriculture: A review", *Sensors,* vol. 18, no. 8, p. 2674, 2018.
[http://dx.doi.org/10.3390/s18082674] [PMID: 30110960]

[7]　F.A. Auat Cheein, and R. Carelli, "Agricultural robotics: Unmanned robotic service units in agricultural tasks", *IEEE Ind. Electron. Mag.,* vol. 7, no. 3, pp. 48-58, 2018.
[http://dx.doi.org/10.1109/MIE.2013.2252957]

[8]　T. Mahindru, "Role of Digital and AI Technologies in Indian Agriculture: Potential and way forward", *Niti Aayog, Government of India, September 2019.* [Accessed: January 3rd 2022].

[9]　J.W. Jones, J.M. Antle, B. Basso, K.J. Boote, R.T. Conant, I. Foster, H.C.J. Godfray, M. Herrero, R.E. Howitt, S. Janssen, B.A. Keating, R. Munoz-Carpena, C.H. Porter, C. Rosenzweig, and T.R. Wheeler, "Toward a new generation of agricultural system data, models, and knowledge products: State of agricultural systems science", *Agric. Syst.,* vol. 155, pp. 269-288, 2017.
[PMID: 28701818]

[10]　United Nation Global Compact, *Digital Agriculture: Feeding the future.* Available From: http://breakthrough.unglobalcompact.org/disruptive-technologies/digital-agriculture/ (Accessed on November 12th 2021).

[11]　Y. Edan, S. Han, and N. Kondo, "Automation in Agriculture", In: *Springer Handbook of Automation,* 2009, pp. 1095-1128.

[12]　R.R. Shamshiri, and K.C. Fatemeh Kalantari, "Advances in greenhouse automation and controlled environment agriculture: A transition to plant factories and urban agriculture", *Int. J. Agric. Biol. Eng.,* vol. 11, no. 1, pp. 1-22, 2018.
[http://dx.doi.org/10.25165/j.ijabe.20181101.3210]

[13]　R.R. Murphy, "Introduction to AI robotics", Available From: Available From: https://mitpress.mit.edu/books/introduction-ai-robotics (Accessed on October 25th 2021).

[14]    S.Y. Nof, Ed., "Handbook of industrial robotics". John Wiley & Sons, 1999.

[15]    Mario Lezoche, Jorge Hernandez, Alemany Diaz Maria del Mar, Hervé Panetto, and Janusz Kacprzyk, "Agri-food 4.0: A survey of the supply chains and technologies for the future agriculture", *Computers in Industry,* vol. 117, p. 103187, 2020.

[16]    Anjin Chang, Jinha Jung, Murilo M. Maeda, Juan Landivar, "Crop height monitoring with digital imagery from Unmanned Aerial System (UAS)",Computers and Electronics in Agriculture, Vol. 141, Pages 232-2372017, doi: ISSN 0168-1699,
[http://dx.doi.org/10.1016/j.compag.2017.07.008]

[17]    S.C. Hassler, and F. Baysal-Gurel, "Unmanned aircraft system (UAS) Technology and applications in agriculture", *Agronomy,* vol. 9, no. 10, p. 618, 2019.
[http://dx.doi.org/10.3390/agronomy9100618]

[18]    R. Hamza, and M. Chtourou, "Apple Ripeness Estimation Using Artificial Neural Network", *International Conference on High Performance Computing & Simulation (HPCS),* 2018pp. 229-234

[19]    S. Lal, S.K. Behera, P.K. Sethy, and A.K. Rath, "Identification and counting of mature apple fruit based on BP feed forward neural network", *Third International Conference on Sensing, Signal Processing and Security (ICSSS),* 2017pp. 361-368

[20]    K. Ang, L. Minn, and S. Leigh, "Computer Vision and Machine Learning for Viticulture Technology", *IEEE Access,* pp. 1-1, 2019.
[http://dx.doi.org/10.1109/ACCESS.2018.2875862]

[21]    P. Skolik, C.L.M. Morais, F.L. Martin, and M.R. McAinsh, "Determination of developmental and ripening stages of whole tomato fruit using portable infrared spectroscopy and Chemometrics", *BMC Plant Biol.,* vol. 19, no. 1, p. 236, 2019.
[http://dx.doi.org/10.1186/s12870-019-1852-5] [PMID: 31164091]

[22]    D. Du, J. Wang, B. Wang, L. Zhu, and X. Hong, "Ripeness prediction of postharvest kiwifruit using a MOS E-nose combined with chemometrics", *Sensors,* vol. 19, no. 2, p. 419, 2019.
[http://dx.doi.org/10.3390/s19020419] [PMID: 30669613]

[23]    S. Cárdenas-Pérez, J. Chanona-Pérez, J.V. Méndez-Méndez, G. Calderón-Domínguez, R. López-Santiago, M.J. Perea-Flores, and I. Arzate-Vázquez, "Evaluation of the ripening stages of apple (Golden Delicious) by means of computer vision system", *Biosyst. Eng.,* pp. 46-58, 2019.
[http://dx.doi.org/10.1016/j.biosystemseng.2017.04.009]

[24]    W. Castro, J. Oblitas, M. De-la-Torre, C. Cotrina, K. Bazán, and H. Avila-George, "Using machine learning techniques and different color spaces for the classification of Cape gooseberry (Physalis peruviana L.) fruits according to ripeness level", *Peer J Preprints,* vol. 7, p. e26691v2, .

[25]    U. Abdulhamid, M.A. Abdulhamid, and S. Daniel, "Detection of soya beans ripeness using image processing techniques and artificial neural network", *Asian J. Phys. Chem. Sci.,* vol. 5, no. 10, 2019.
[http://dx.doi.org/9734/ajopacs/2018/39653]

[26]    O.S. Taghadomi-Saberi, E-D. Mahmoud, and K. Faraji-Mahyari, "Determination of cherry color parameters during ripening by artificial neural network assisted image processing technique", *J. Agri. Sci. Tech.,* vol. 17, pp. 589-600, 2015.

[27]    F. Bader, and S. Rahimifard, "Challenges for industrial robot applications in food manufacturing", *Proc. 2nd Inte. Symp. Comp. Sci. Intel. Cont.,* 2018p. 37 Stockholm, Sweden

[28]    N. Chandra, B. Tudu, and C. Koley, "A Machine Vision-Based Maturity Prediction System for Sorting of Harvested Mangoes", *Instr. Measu., IEEE Trans.,* vol. 63, pp. 1722-1730, 2014.
[http://dx.doi.org/10.1109/TIM.2014.2299527]

[29]    S. Mohammad, and J. Bansal, "Introduction to computer vision and machine learning applications in agriculture", In: *Algor. Intel. Sys.,* 2021, pp. 1-8.

[30]    X.C. Ting Yuan, "Information acquisition of cucumber fruit in greenhouse environment based on nearing infrared image", *Inf. Process. Agric.,* vol. 7, no. 1, pp. 1-19, 2020.
[http://dx.doi.org/10.1016/j.inpa.2019.09.006]

[31]    Q. Zhang, S. Chen, and T. Yu, "Cherry recognition in natural environment based on the vision of picking robot", *IOP Conf. Ser. Earth Environ. Sci.,* vol. 61, no. 1, 2017.
[http://dx.doi.org/10.1088/1755-1315/61/1/012021]

CHAPTER 5

# Field Prevention System from Wild Animals

**Mayank Patel[1,\*], Latif Khan[1], Saurabh Srivastava[1]** and **Harshita Jain[1]**

[1] *Geetanjali Institute of Technical Studies, Udaipur, Rajasthan, India*

**Abstract:** Preventing wild animal attacks in fields is a highly challenging task for farmers and field holders, especially during nighttime. Continuous monitoring is difficult to maintain consistently. Therefore, we have designed an Intrusion Detection System based on the Internet of Things (IoT). Our system utilizes the ESP8266 as its central component, allowing for the implementation of an automated solution to repel animals from fields without human intervention. Various devices, such as hooters, flashlights with day-night vision cameras, and AI algorithms, are incorporated to detect and differentiate animals from humans. Additionally, mobile applications provide a convenient means to remotely monitor the system's actions from home.

**Keywords:** IoT technology, Intrusion detection, Field rrotection, ESP8266, Artificial intelligence, Mobile application, Day-night vision camera.

## INTRODUCTION

The agriculture sector is the largest sector in India, encompassing a wide range of activities. Unfortunately, it is also one of the most severely affected areas, with farmers encountering numerous difficulties and challenges in their livelihoods and cultivation [1]. The challenges faced by farmers are extensive, impacting approximately 40% of the farming sector nationwide.

Given that the agricultural sector is crucial to the global economy, it holds significant importance for India as well [2]. Therefore, it becomes imperative to address the issues faced by farmers and find effective solutions to combat these problems. Field preservation is a critical aspect that needs to be prioritized in order to safeguard the sector and ensure its sustainable growth.

Farmers have emerged as the true backbone of India's development, with agriculture covering the largest sector in the country. However, Indian farmers face numerous day-to-day challenges and problems, which vary based on various factors [3].

---

[\*] **Corresponding author Mayank Patel:** Geetanjali Institute of Technical Studies, Udaipur, India;
E-mail: mayank999_udaipur@yahoo.com

Praveen Kumar Shukla & Tushar Kanti Bera (Eds.)

One major challenge is the encroachment of wild animals into agricultural areas due to the loss of habitat and increased urbanization. As cities expand and forest areas shrink, animals are forced to search for prey in agricultural fields. This situation, although unusual, has become a common concern for farmers, causing worry and affecting cultivation. Animals now directly enter fields, posing a threat to both humans and their food. This sudden aggression by animals towards humans necessitates the adoption of eco-friendly methods to keep them away from fields and prevent their entry.

Despite being the most economically significant sector in India, the agricultural sector is plagued by the issue of human-animal conflict. Wild animals entering fields, destroying crops, and disrupting the livelihoods of small-scale farmers are major concerns. Traditional methods still in use today can be dangerous for all parties involved. Given that field owners cannot guard their land 24/7, there is an urgent need to implement an eco-friendly system that can effectively deter animals from entering fields and ensure their safe escape.

## LITERATURE REVIEW

The current system still in use today relies on electrical fencing surrounding the field, which delivers electrical shocks to animals, posing a danger Fig. (1). Various methods are currently employed, including the use of flambeaus, firecrackers, egg repellents, fish smell, *etc* [4]. Some attempts have been made to address the issue using motion sensors. However, the sensitivity of these sensors can present a barrier to accurately detecting animal's entry into the field.

**Fig. (1).** Figures showing the existing methods.

Many individuals and organizations resort to trial and error, as they lack technological and eco-friendly solutions to solve such problems [5]. The most prevalent method currently employed is the electric fence, consisting of a wire perimeter charged with electricity [6]. This method serves as a deterrent for both animals and humans. The electric charge can be adjusted by varying the potential, resulting in effects ranging from discomfort to death. Electric fences are commonly used for agricultural fencing and animal control purposes. They work by creating an electrical circuit when touched by a person or animal [7]. A power energizer, a key component, converts power into brief high-voltage pulses. One terminal of the power energizer releases these pulses along a connected bare wire approximately once per second. The other terminal is connected to a metal rod, known as a ground or earth rod, which is implanted in the earth [8]. When a person or animal touches both the wire and the earth during a pulse, an electrical circuit is completed, resulting in an electric shock.

In the aforementioned pictures, we observed the use of flambeaus and firecrackers by farmers in the agricultural sector as a means to deter animals. However, it is important to note that this method carries a risk, as it could potentially result in an unexpected counter-attack [9].

Many farmers resort to electrical fencing, which involves the installation of simple wires around the field. The fencing is powered by electricity, and when an animal touches or comes into contact with it, they receive a sudden electric shock, posing a potential threat to their life [10].

## PROPOSED INNOVATION SYSTEM

When animals enter or attempt to enter the field, the continuity of the two-level laser fencing is disrupted, triggering the capture of images by a day-night vision camera. These images are then sent to a server. Simultaneously, irritating sounds are emitted from a sound-producing device, and high-intensity flashlights are directed towards the location of the animal Fig. (2). To notify the field owner and provide intrusion information, alert notifications are sent *via* messages. The entire data can be accessed through a hybrid application platform that receives updates from the server.

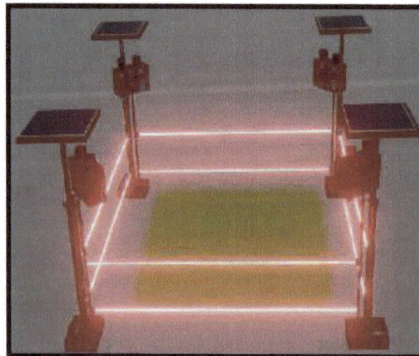

**Fig. (2).** The image of proposed System.

The entire system operates on solar energy, making it a solar-based system that harnesses the power of the sun. Additionally, an underground fencing mechanism is implemented to prevent underground attacks. Once a wild animal is detected, the entire system is activated. The system also monitors the temperature and humidity surrounding the field, providing alerts in the event of a fire. The detection of wild animals is carried out through the utilization of Artificial Intelligence. Consequently, when the application is opened, the user can view classified images of animals alongside their respective identifications.

The image provided above depicts a schematic representation of the two-level laser fencing system. During nighttime, the laser beams will be visible, and when an animal crosses the lasers, the system will detect the intrusion. The two-level laser fencing is designed to target both tall and short animals effectively. Upon detection, the system will activate irritating sounds and direct high-intensity flashlights toward the intruded area. Additionally, a day-night vision camera captures an image of the animal and sends it to the server. The server then updates the mobile application with an alert SMS, notifying the farmer of the intrusion. By accessing the mobile application, the farmer will be able to identify the type of animal involved.

In image classification, Convolutional Neural Networks (CNNs) play a significant role. CNNs employ different numbers of computational layers for feature learning from input images, depending on the specific visual task [11].

**Regular CNN**

A regular Convolutional Neural Network (CNN) is commonly used for computer vision tasks like image classification and object detection. CNN is known as a deep learning algorithm and consists of two main components: feature learning and classification [12]. In comparing Fast R-CNN, Faster R-CNN, and their proposed method, an average F-measure score of 82.1% was achieved for animal species detection [13].

For feature learning, multiple hidden layers (consisting of convolutional and pooling layers) perform convolution and pooling operations on the input data. This process generates feature maps with learned weights over the input image and local regions that match the filter size (receptive field) [14, 15]. These layers use a non-linear activation function called Rectified Linear Unit (ReLU) and employ backpropagation to extract object features from the image, irrespective of their location [16]. Multiple filters are applied, resulting in multiple feature maps [17].

Pooling is a downsampling operation applied to the output of a convolutional layer. It reduces redundant information, enabling the extraction of significant features related to objects in the input image. Two common pooling methods are average pooling and max pooling, which calculate the average value or extract the maximum value from each region on the feature map, respectively [18, 19].

In Fig. (3), the Intrusion System is currently inactive, and no intrusion has occurred yet. Let's assume animals have entered and broken the laser fencing. The camera integrated into the control unit will move towards the intrusion direction and capture an image. Simultaneously, the hooters and flashlights will activate their functions. Once the animal is detected, classification will be performed to identify its species, and the mobile application will be updated with an alert SMS.

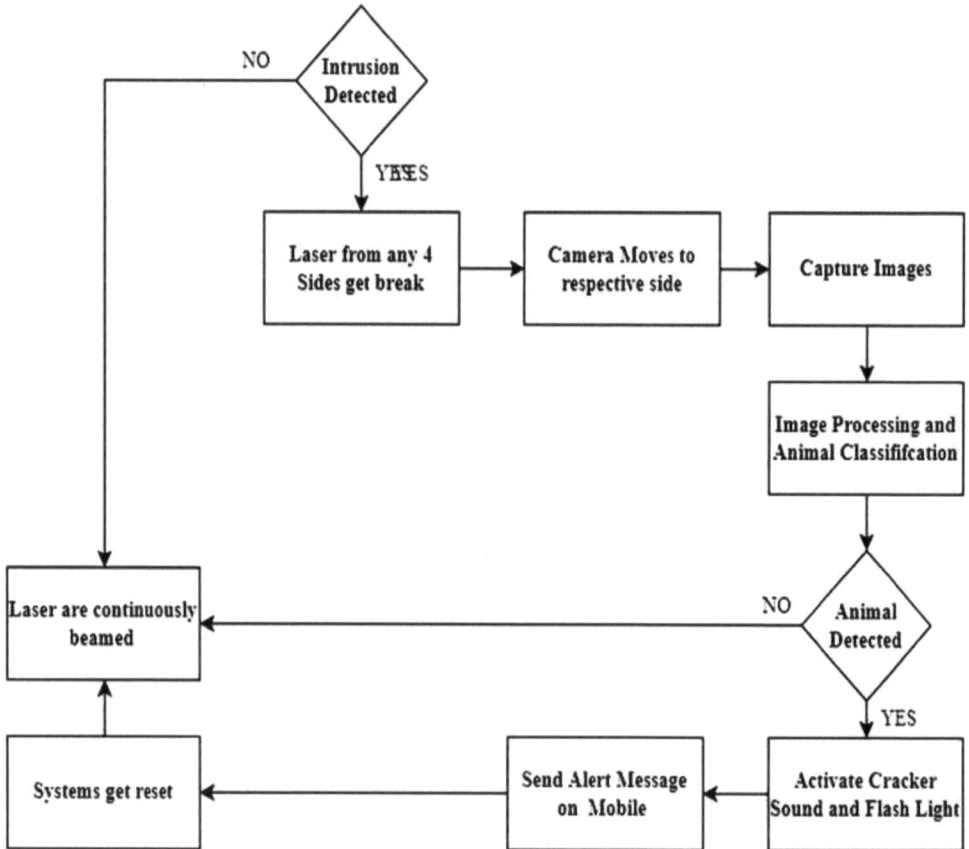

**Fig. (3).** Workflow of proposed system.

## FLOWCHART

The workflow of the proposed system is shown in Fig. (**3**).

## Algorithm:

## START:

Step1:- Intrusion detection using long-range Laser and Photodiode Sensors.

Step2:- If "Laser from any of the 4 sides breaks", then, Day-Night Vision Surveillance Camera moves to the respective intruded side. Go to step-3. Else Repeat from Step-1.

Step-3:- Capture the Image.

Step-4:- Image Processing and Animal classification.

Step-5:- If "Animal detected", then, Go to step-6. Else No Action. Repeat from step-1.

Step-6:- Activate different loud irritating sounds through Hooters and high intensity flashlights.

Step-7:- Update Server and hybrid Mobile Application.

Step-8:- Send Alert Message to farmers mobile. End.

## SYSTEM REQUIREMENTS

## OPERATING SYSTEM-

- Windows XP/Vista/7/8.
- Android.

## SOFTWARE REQUIREMENTS

- **Thing Speak Server:** It is a web cloud server that is used for IoT Technology.
- **Android Studio:** It is used for designing applications.

## HARDWARE REQUIREMENTS

- RAM: - 4 GB (recommended).
- HDD: - 40 GB (recommended).
- Processor: - I3 or any other.
- Internet: - Independent LAN/WAN for the company.

- IoT Components.
- Sensors.
- Processing Boards.

## DESIGN AND IMPLEMENTATION CONSTRAINTS

- For designing and implementing this project, we use different IoT components which are:

### Sensors

1. Laser Receiver and Transmitter Fig. (**4**). These components operate on a binary concept, where they function as digital sensors and activate only when an intrusion occurs. The receiver does not receive light if there is an obstacle disrupting the continuity of the transmitted light [20]. The system operates on the principle that a value of 0 on the server indicates safety, and when the laser is interrupted, it immediately changes to 1, indicating an intrusion has occurred [21].

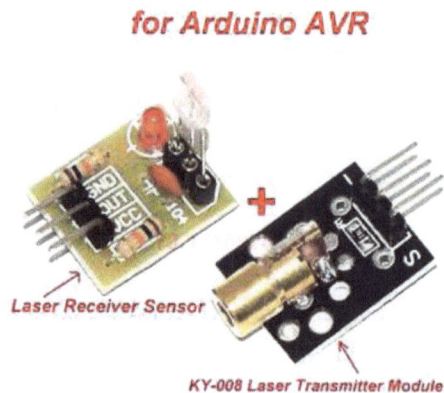

**Fig. (4).** Laser Rx and Tx.

2. Temperature and Humidity Sensor (Fig. **5**) – In the system, we will use a temperature and humidity sensor modelled at the DHT11 sensor for determining temperature and humidity [22].

**Fig. (5).** Temperature Humidity detecting sensor.

3. Hooters/Speakers (Fig. **6**) for making loud irritating noises and different kinds of sounds such that animals did not get into the habitat of it on regular arrival.

**Fig. (6).** Hooters and Speakers used in the system.

4. Solar Panels for utilizing nature's resources instead of electricity (Fig. **7**).

**Fig. (7).** Solar Panel for green energy.

5. Flashlights (Fig. **8**).

**Fig. (8).** High Intensity Flashlights.

6. Day-Night Vision Camera (Fig. **9**).

**Fig. (9).** Day-Night vision camera for capturing intruded image.

## Boards

**1.** ESP8266 - ESP8266 stands for Espressif Systems' Modules. It is a low-cost Wi-Fi microchip that features a full IP/TCP stack and microcontroller capabilities [23]. This type of firmware utilizes the Lua scripting language (Fig. **10**), which has been introduced by Espressif Systems [24]. The recommended voltage range for the module is around 3.3V, with a voltage range between 2.5V (minimum) and 3.6V (maximum) [25].

**Fig. (10).** ESP8266 microchip.

**2.** ESP32 CAM Board (Fig. **11**) is a Wi-Fi low-cost camera module for sharing the image to the mobile application through which the farmers will be benefitted [26].

**Fig. (11).** ESP32 CAM Module for communication for image transfer.

## Others

Other additional components include.

1. PCB.

2. LED Fig. (**12**).

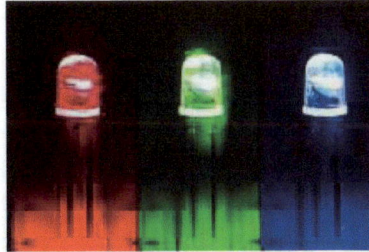

**Fig. (12).** LED.

3. Soldering station.

4. Servo Motor.

5. Battery.

6. Solar charge controller.

7. Resistors Fig. (**13**),

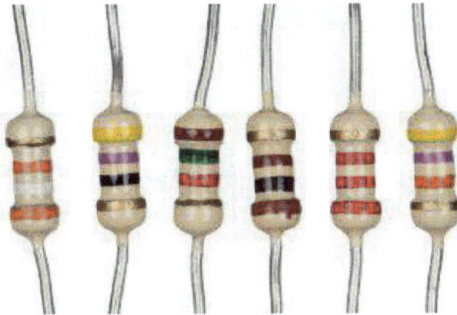

**Fig. (13).** Resistors.

8. Pins.

9. Multi-meters.

10. Connecting wires Fig. (**14**).

**Fig. (14).** Connecting wires.

11. Long range wires.

## BLOCK DIAGRAM

The block diagram Fig. (**15**). Depicts the characteristics of the sensors and the boards connected with each other. Here, for the pole, we have 4 optical sensors preferably a digital sensor also known as a laser receiver. We are currently working on solar fencing and a battery mechanism for the system and Mobile Application would be helpful for the information.

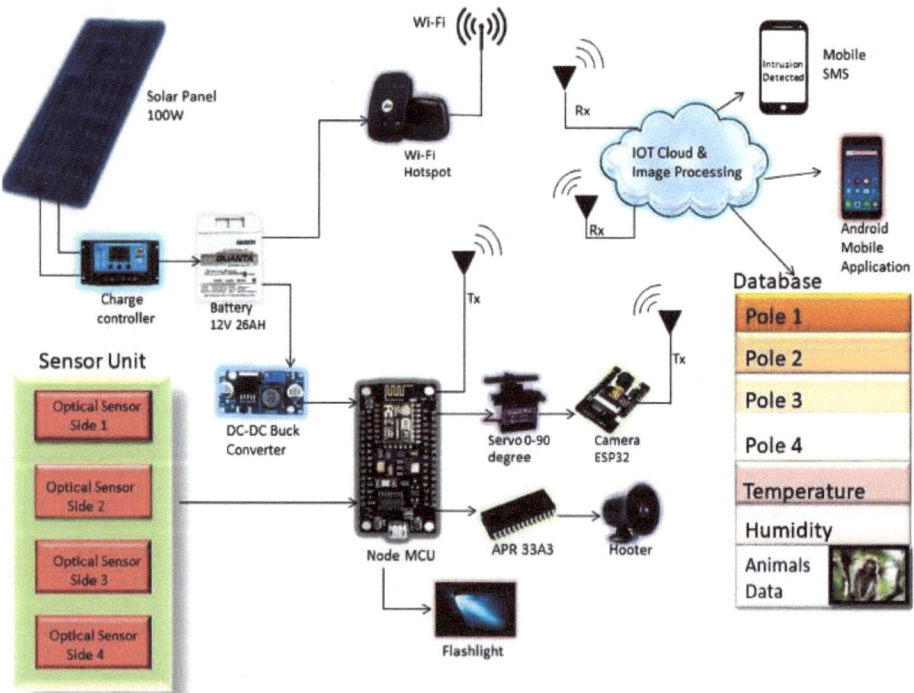

**Fig. (15).** Block Diagram.

The deep functioning of the system involves the synchronization of various essential sensors and the routing of connections and their connectivity. The system operates using a 100W 12V solar panel that harnesses solar energy for power. The solar charge controller is responsible for managing the charging process and monitoring energy information, ensuring proper charge levels to prevent undercharging or overcharging. The system is powered by a 12V 26AH battery, which provides power during night time when solar energy is not available.

Each pole is equipped with laser sensors that operate on a 0 and 1 concept. A reading of 0 indicates that no intrusion has occurred, while a reading of 1 indicates an intrusion. The Node MCU, also known as ESP8266, acts as a WiFi chip and oversees the workflow of the system. It is responsible for updating the application through APIs, but the presence of a WiFi hotspot is not mandatory for the system to function properly.

The system is designed to be stable and perform excellently in transmitting messages and data. The devices integrated into the system's motherboard, including high-intensity flashlights, speakers for playing loud noises, and a surveillance camera with day-night vision capabilities, ensure smooth operation and efficient performance.

## HARDWARE RESULT

As we have seen ThingSpeak Server charts of 4 poles each, each pole carries its data through Fig. (**16**) which would be helpful to know whether the system is working properly or not [27].

Now we have 4 charts that are Temperature, humidity, Times of Intrusion, and their occurrence in the field, and the digital presence of the animal in the field.

We have received an SMS stating that the intrusion has occurred in the field in both Hindi and English language for better user experience Fig. (**17**).

As observed in Fig. (**18**), the mobile application is utilized. On the homepage, the Intrusion system is shown as "armed away," indicating that no intrusion has occurred yet [28]. Additionally, the application provides weather information, which is beneficial for farmers to determine irrigation needs for their land and also serves as a tool for fire alertness in the field during emergencies.

## Readings at Pole 1

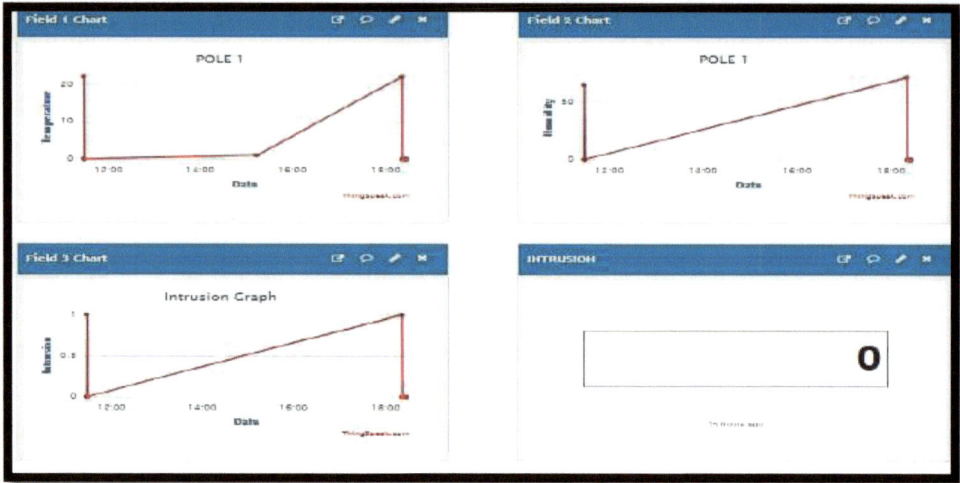

## Readings at Pole 2

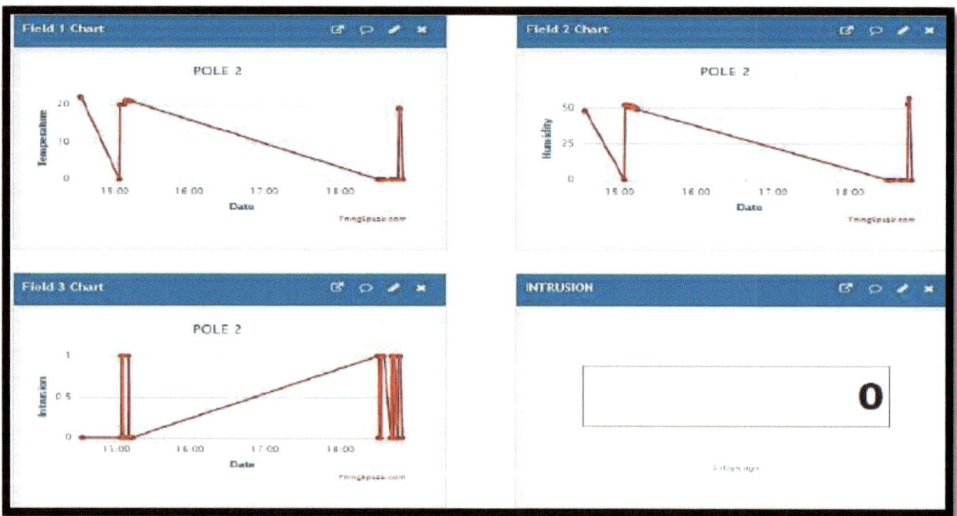

*(Fig. 16) contd.....*

## Readings at Pole 3

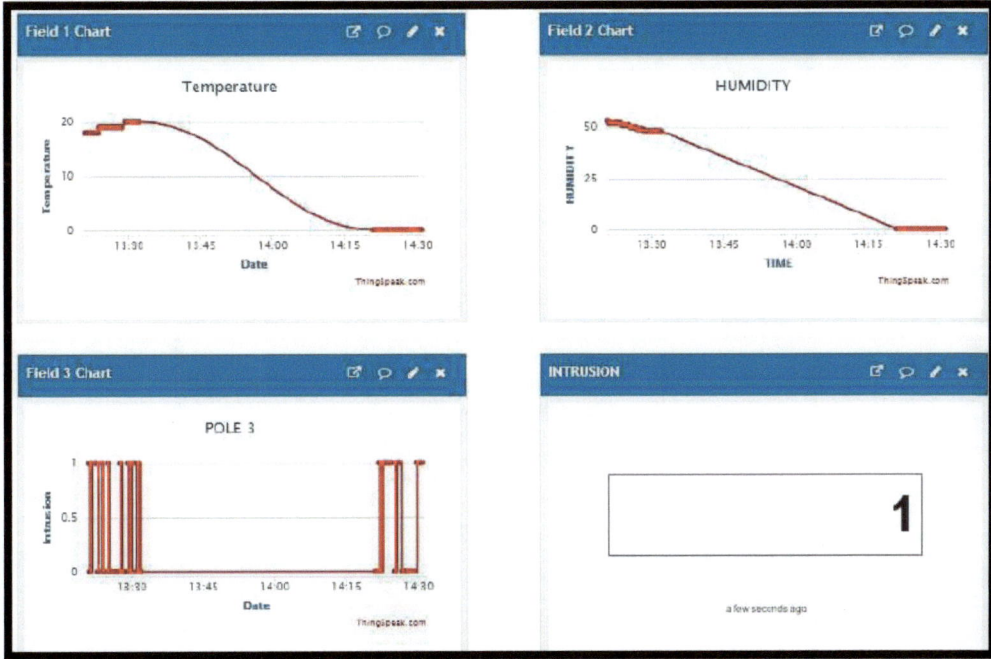

## Readings at Pole 4

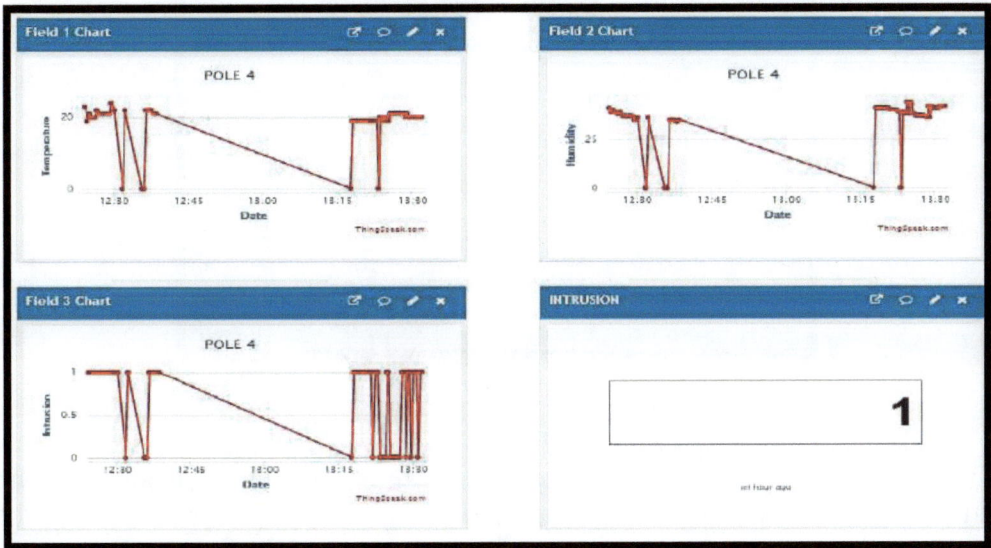

**Fig. (16).** Figure shows the output on the ThingSpeak Server when the animal enters.

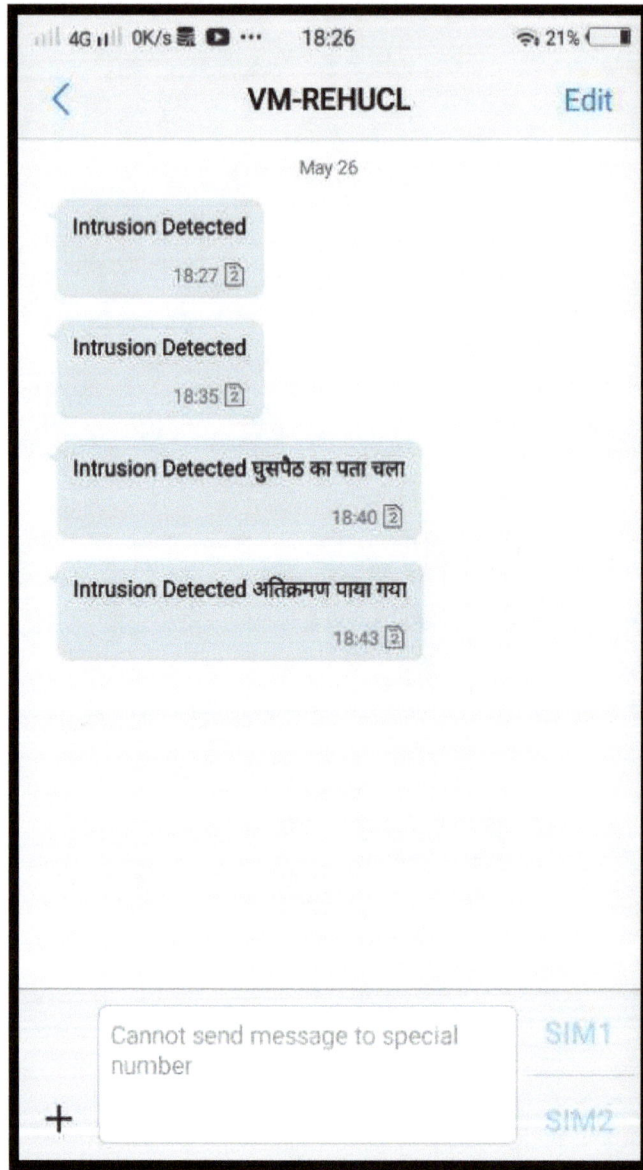

**Fig. (17).** Screenshot of message on phone.

When the Intrusion System is triggered, an alarm is activated, indicating that an intrusion has occurred. In this case, the application prompts the user to check the image portal to identify the type of animal that has entered the field [29].

Field safe             Weather Information

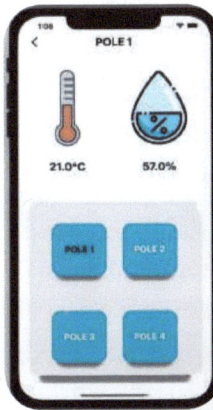

Temperature and Humidity        Field Attacked

Intrusion Image sample

**Fig. (18).** Mobile Application for end users with data and image.

## CONCLUSION

We have developed a system to address the issue of man-animal conflict and protect fields from wild animal intrusion. This solution is eco-friendly and prevents animals from entering the fields by employing methods such as playing irritating loud noises and using high-intensity flashlights to deter them. The system also sends alerts to the owner regarding intrusions and fire incidents [30]. To promote energy efficiency, the entire system operates on solar panels.

Artificial Intelligence plays a crucial role in classifying animals based on their pictures, enhancing the functionality of the application. The system efficiently preserves and safeguards fields in an eco-friendly manner, eliminating the need for human intervention. It is a cost-effective solution that ensures no harm to animal life and minimizes losses in cultivation [31].

The intelligent sensors integrated into the system utilize IoT technology, utilizing data and sensor inputs. With the assistance of AI-based classification, the system can continuously improve the application's performance and enhance the user experience. In the future, this technology could even help in detecting the movement of wild animals crossing forest boundaries.

## ACKNOWLEDGMENTS

We hereby thank our Institute, Geetanjali Institute of Technical Studies, Udaipur for motivating all the authors to contribute to this work.

## REFERENCES

[1]    S.I. Sherly, M. Haripushpam, B.I. Rinie, and D.J. Devi, "Wildlife intrusion and early fire detection system using IoT", *Inter. J. Innov. Res. Comp. Commun. Engin.,* vol. 6, no. 2, pp. 1837-1840, 2018.

[2]    V.M. Kakaderly, "Design and implementation of an advanced security system for farm protection from wild animals", *Global Research and Development Journal for Engineering,* vol. 4, no. 3, pp. 14-18, 2019.

[3]    P. Navaneetha, R.R. Devi, S. Vennila, P. Manikandan, and S. Saravanan, "IOT based crop protection system against birds and wild animal attacks", *Inter. J. Innov. Res. Techno.,* vol. 6, no. 11, pp. 138-143, 2020.

[4]    S. Santhiya, Y. Dhamodharan, N.E. Kavi Priya, C.S. Santhosh, and M. Surekha, "A smart farmland using raspberry pi crop prevention and animal intrusion detection system", *Inter. Res. J. Engin. Techn.,* vol. 5, no. 3, pp. 1054-1058, 2018.

[5]    S.J. Sugumar, and R. Jayaparvathy, "An improved real time image detection system for elephant intrusion along the forest border areas", *Sci. Wo. J.,* vol. 2014, pp. 1-10, 2014.
[http://dx.doi.org/10.1155/2014/393958] [PMID: 24574886]

[6]    V. B, R. B, S. R, S. S, and P. G, "Animal detection system in farm areas", *Int. J. Adv. Res. Comput. Commun. Eng.,* vol. 6, no. 3, pp. 587-591, 2017.
[http://dx.doi.org/10.17148/IJARCCE.2017.63137]

[7]     N. Andavarapul, and V.K. Vatsavayi, "Wild-animal recognition in agriculture farms using w-cohog for agro-security", *Int J. Computational Intelligence Research,* vol. 13, no. 9, pp. 2247-2257, 2017.

[8]     P.V. Rao, S.R. Krishna, and M.S. Reddy, "A smart crop protection against animals attack", *Inter. J. Sci. Res. Review,* vol. 8, no. 6, pp. 407-410, 2019.

[9]     S. Vidhya, T. J. Vishwashankar, K. Akshaya, A. Premdas, and R. Rohith, "Smart crop protection using deep learning approach", *Int. J. Inno. Tech. Expl. Eng.,* vol. 8, no. 8, pp. 301-305, 2019.

[10]    B. Krishnamurthy, M. Divya, S. Abhishek, and H.A. Shashank, "Solar fencing unit and alarm for animal entry prevention", *Int. J. Latest Engi. Res. Appl.,* vol. 02, no. 05, pp. 128-135, 2017.

[11]    S. Pandey, and S.B. Bajracharya, "Crop protection and its effectiveness against wildlife: A case study of two villages of shivapuri national park, nepal", *Nepal J. Sci. Technol.,* vol. 16, no. 1, pp. 1-10, 2016.
[http://dx.doi.org/10.3126/njst.v16i1.14352]

[12]    S. Srivastava, M. D. Jain, H. Jain, A. Soni, G. Anand, H. Jain, J. Suthar, L Khan, and M Patel, "Self-intrusion detection system for protection of agricultural fields against wild animals", *Int. J. Mod. Agricul.,* vol. 10, no. 2, pp. 2686-2691.

[13]    S.K. Chauhan, A. Sharma, and A. Kaur, "Animal intrusion detection and prevention system", *Inter. J. Comp. Organiz. Trends,* vol. 11, no. 2, pp. 25-28, 2021.
[http://dx.doi.org/10.14445/22492593/IJCOT-V11I2P308]

[14]    P.U. Maheshwari, and A.R. Rajan, "Animal intrusion detection system using wireless sensor networks", *Inter. J. Advanced Research in Biology Engineering Science and Technology,* vol. 2, pp. 1369-1373, 2016.

[15]    M. Baharuddin, A. Zayegh, and R.K. Begg, "Ultrasonic and infrared sensors performance in a wireless obstacle detection system", *1st International Conference on Artificial Intelligence, Modelling and Simulation,* 2013pp. 439-444 Kota Kinabalu, Malaysia

[16]    S Jeevitha, and V. Kumar, "A review of animal intrusion detection system", *Int. J. Eng. Res. Technol.,* vol. V9, no. 5, pp. 5595-5587, 2020. [IJERT].
[http://dx.doi.org/10.17577/IJERTV9IS050351]

[17]    V. Mitra, C.J. Waang, and G. Edwards, "Neural network for lidar detection of fish", *Proceedings of the International Joint Conference in Neural Networks,* vol. vol. 2, 2003pp. 1001-1006
[http://dx.doi.org/10.1109/IJCNN.2003.1223827]

[18]    T. Burkhardt, and J. Calic, "Real-time face detection and tracking of animals", *8th Seminar on Neural Network Application in Electrical Engineering,* 2006pp. 27-32 Belgrade, Serbia
[http://dx.doi.org/10.1109/NEUREL.2006.341167]

[19]    H. Kaiming, "Deep residual learning for image recognition", *IEEE Conf. Comp. Visi. Pat. Recogn.,* 2016pp. 770-778 Las Vegas, NV, USA
[http://dx.doi.org/10.1109/CVPR.2016.90]

[20]    S. Kumar Roy, A. Roy, and S. Misra, "AID: A prototype for agricultural intrusion detection using wireless sensor network", *IEEE International Conference on Communications (ICC),* 2015pp. 7059-7064 London, UK

[21]    S.R. Thanuja, B. Lekhana, S. Apoorva, R. Bhavana, and L. Bhaskar, "Crop protection and animal intrusion detection system", *J. Emerg. Technol. Innov. Res.,* vol. 7, no. 7, 2020.

[22]    R. Patil, J. Gayathri, K. Ashwini, and K.K. Gururaj, "Protection of crops and usage of rain water using IoT", *Inter. J. Advanced Research in Electrical, Electronics and Instrumentation Engineering,* vol. 7, no. 6, pp. 3017-3022, 2018.

[23]    U.K. Divya, and M. Praveen, "IOT- Based wild animal intrusion detection system", *Int. J. Recent Innov. Trends Comput. Commun.,* vol. 6, no. 7, pp. 6-8, 2018.

[24]    J.C. Nascimento, and J.S. Marques, "Performance evaluation of object detection algorithms for video surveillance", *IEEE Trans. Multimed.,* vol. 8, no. 4, pp. 761-774, 2006.
[http://dx.doi.org/10.1109/TMM.2006.876287]

[25]    R. Balathandapani, D. Boopathi, S. Jotheeshwaran, and G. Arundeva, "Automatic rain water and crop saving system using embedded technology", *Inter. J. Scientific Engineering and Technology Research,* vol. 4, no. 3, pp. 314-317, 2015.

[26]    C. Peijiang, "Moving object detection based on background extraction", *International Symposium on Computer Network and Multimedia Technology,* 2006pp. 1-4 Wuhan, China
[http://dx.doi.org/10.1109/CNMT.2009.5374629]

[27]    T. Burghardt, and J. Calic, "Real-time face detection and tracking of animals", *8th Seminar on Neural Network Applications in Electrical Engineering,* 2006pp. 27-32 Belgrade, Serbia
[http://dx.doi.org/10.1109/NEUREL.2006.341167]

[28]    M.F. Thorpe, A. Delorme, and S.T.C. Marlot, "A limit to the speed processing in ultra-rapid visual categorization of novel natural scene", *Cogn. Neurosci.,* pp. 171-180, 2003.

[29]    P. Goutham Goud, N. Suresh, E. Surendhar, G. Goutham, and V.M. Kiran, "Rain Sensor automatically controlled dryingshed for crop yield farms I International Research", *J. Eng. Technol.,* vol. 4, no. 7, pp. 573-575, 2017.

[30]    N. Datta, "Automatic tracking and alarm system for eradication of wild life injury and mortality", *2nd International Conference on Advances in Computing, Communication, & Automation (ICACCA) (Fall),* 2016pp. 1-4 Bareilly, India
[http://dx.doi.org/10.1109/ICACCAF.2016.7748972]

[31]    S.U. Sharma, and D.J. Shah, "A practical animal detection and collision avoidance system using computer vision technique", *IEEE Access,* vol. 5, pp. 347-358, 2017.
[http://dx.doi.org/10.1109/ACCESS.2016.2642981]

**CHAPTER 6**

# Weather Forecasting using Machine Learning for Smart Farming

**Rajan Prasad**[1,*] and **Praveen Kumar Shukla**[1]

[1] *Artificial Intelligence Research Center, Department of Computer Science & Engineering, School of Engineering, Babu Banarasi Das University, Lucknow, India*

**Abstract:** Weather forecast is of prime attention of the researchers working in the smart agriculture domain. In India, approximately 55% of the total crops are dependent on weather (monsoon season). An accurate weather forecast model requires abundant data to get the most accurate predictions. However, the weather forecast is a key area of research and is always challenging from historical data. Hence, the current system used for weather forecasting is an amalgamation of forecasting models, opinions, and information trends, and specific patterns. This work presents the application of the linear regression model and polynomial regression model for weather forecasting; like a scheme to forecast rainfall, and precipitation using historical weather data. The sample weather dataset covers 75 districts of Uttar Pradesh state which is received from the Indian Meteorological Department (IMD). Furthermore, analysing the impact of forecasts with different parameters is realized over six major crops Triticum (biological name of wheat), Gram, Barley, Mustard, Sugarcane, and Maize of Uttar Pradesh State. The main objective of the state-of-the-art is efficient crop management and passing the appropriate message to farmers to make suitable decisions as per the weather conditions.

**Keywords:** Future Farming, Linear Regression, Machine Learning, Weather forecast.

## INTRODUCTION

Accurate weather forecasting is a critical task depending on several parameters. Many advanced techniques like machine learning, deep learning, *etc.* are being used to design and develop weather forecasting systems for agriculture purposes with higher precision and accuracy [1, 2].

Uttar Pradesh is the 4th largest  state in India and has the  highest  population.  It is situated in  the predominantly subtropical region. The  weather  depends  on

---

* **Corresponding author Rajan Prasad:** Artificial Intelligence Research Center, Department of Computer Science & Engineering, School of Engineering, Babu Banarasi Das University, Lucknow, India; E-mail: rajan18781@gmail.com

**Praveen Kumar Shukla & Tushar Kanti Bera (Eds.)**

elevation; the average temperature varies from 12.5°C -17.5°C in January and 27.5°C–32.5°C from May to June. The distribution of rainfall may vary from 1000 to 2000 mm, most of the rainfall occurs during the southwest monsoon starting from mid-June to September months when most of the crops depend on rainfall. The annual rainfall affects the profit or loss of farm products. Crop loss may be minimized by taking precautions based on accurate weather forecasts. So, getting essential information about weather like temperature, rainfall, or natural disasters such as droughts, flood storms, *etc.* helps the farmers to manage their jobs.

Machine learning techniques usually consider historical data to forecast different parameters of weather datasets. The performance of the model depends on two factors; first, whether it is short-term or long-term predictions, second spatial scale, and global scale predictions. The long-term and global scale predictions have better performance rather than short-term and spatial scales. Many applications have used short-term temperature predictions focused on support vector machine (SVM) [1], Multilayer Perceptron (MLP) [2], and Random Forest (RF) [3]. Artificial Neural Networks (ANN) (inspired by biological neurons) [4] are widely used in developing such systems. ANNs are based on the theory of enormous interconnection and the parallel computing model of a biological system. It is a mathematical model that is capable of solving complex engineering problems and is useful in identifying the typical non-linear relationships between input variables to target variables in the dataset. The relationship between input and output variables can be mathematically formulated as given below [5].

$$Y = f(X^n) \tag{1}$$

Where, $X^n$ denotes the input vector of n-dimensional such as variables $X_1, \ldots X_i, \ldots X_n$ and Y denotes the output vector. The function $f(.)$ represented in equation (1) cannot be exposed explicitly from the model; rather, it is represented by the network parameters.

The structure of ANN consists of three layers *i.e.*, input, hidden, and output layers. There is no relationship between the input and output data layer. The hidden layers are often connected by bias, activation functions, and weight matrices. Fig. **(1)** shows the typical structure of ANN.

The complexity of ANN depends on the number of hidden layers. It also depends on the nature of the dataset such as data size, data types *etc.* but there is no standard rule that defines the number of hidden layers in a neural network architecture.

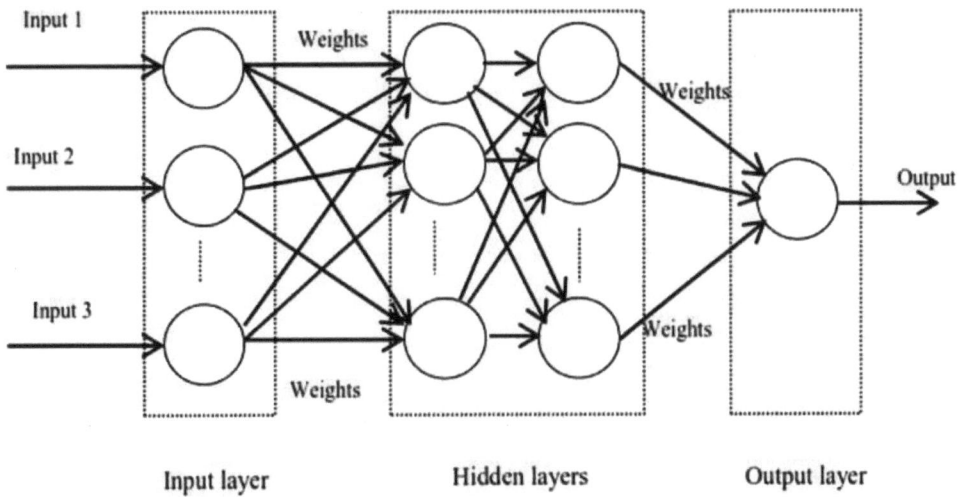

**Fig. (1).** Structure of Artificial Neural Network.

Typically, the gradient descent algorithm is used for training of neural network model and the back-propagation algorithm is used for synaptic weights modification based on the calculation of loss functions. The loss function can be defined as the difference between actual input and predicted output. It also depends on the type of problems. Different loss functions will be applied for the different problems because it has direct significance to the activation function used for calculating the output at the output layer.

The back propagation algorithm [6] is a supervised learning algorithm [7] that allows the update in the synaptic weights of the network by optimising the loss function. The calculated gradients of the loss function are fed to the previous layer of the network. It means that the values of gradients tend to reduce as the number of iterations moves backward in the previous layer of the network.

As a result, the learning rate of the earlier layer neurons slows down and the training period also increases along with a decrease in the accuracy of the model. This is known as the vanishing gradients problem, which is a very difficult problem when training neural networks by the backpropagation algorithm. Moreover, the ANN model has a significant limitation, for processing sequential data such as weather forecasts, finding the word for sentence completion problems, or computing time series data. The recurrent neural network (RNN) [8] is a type of ANN that is capable of handling the above discussed problems. The recurrent neural network was introduced in 1980 [9]. The structure of a recurrent neural network is like chains of modules, these modules act as a memory unit that stores information of previous steps. RNN consists of a feedback loop that permits the network to just accept a chain of inputs. It means the output generated from the step is fed back into a network that impacts the output of the step similar process repeats in all subsequent steps.

Attention is paid to weather forecasting using the Long-Short Term Memory (LSTM) network which is one of the forms of RNN. LSTM networks have been successfully applied to solve various types of forecasting problems such as traffic forecasting [10], electricity load forecasting [11], solar irradiance forecasting [12], and many more.

The paper is divided into five sections. Section 2 represents the literature review; section 3 represents the proposed methods for weather forecasting; section 4 shows the experimental results and discussion, and Section 5 represents the conclusion and future research directions.

## LITERATURE REVIEW

In a study [13], a model of weather forecasting is proposed focusing on data filtering methods for handling noisy data. Fuzzy logic concepts are used for temperature forecasting and prominent results in terms of accuracy are found [14]. The wind speed is estimated using multivariate regression. The proposed model achieved prominent results [15]. A linear regression model has been implemented and compares the accuracy with time series for weather forecasts [16]. A passenger satisfaction model based on a linear regression model is proposed [17]. In [18], a linear time series model is proposed and implemented to predict the temperature of Jaipur city. A high-precision information model is proposed [19] where a study based on Self Organising Map and Latent Dirichlet Allocation model is done. This model is useful for weather prediction and crop management. The dataset used by the model contains more than 1000 records with different types of features such as soil type, crop type, humidity, temperature, and rainfall. Data mining algorithms were applied to find suitable environments for different types of crops. The proposed model achieved 80.09% accuracy. The experimental results showed improved accuracy as compared to existing models.

A supervised learning-based SOM model for rainfall analysis related to synoptic circulation is proposed and implemented [20]. The authors investigated two different rainfall zones in South Africa. The model has the ability to assess the SOM and match the synoptic movers of observed rainfall records, effectively [20]. A model for weather forecasting using data mining tools and historical data is proposed [21]. The authors used a decision tree (DT) algorithm to predict rainfall. The information gained is used for splitting criteria. Validation methods were used for checking the performance of the model. The model has achieved 100% accuracy in the training phase and 80.67% in the testing phase.

An efficient system for the agriculture field including weather forecasting, drought management and precision irrigation forecasting that enhanced the yields is developed [22]. The framework of the model is divided into four layers with strong cohesion and low coupling implemented with hybrid programming. Drought forecasts and water requirements are the modules of the framework coupled with the flow of the data. Due to the complexity of the domain such as natural disasters, the algorithms of the system are enhanced time-to-time. Meanwhile, the multisource of the natural disaster system needs to add/remove functionality that impacts assessment.

A deep learning model for predicting rainfall using multilayer perceptron and auto-encoder neural networks is implemented [23]. The accuracy of the model is measured in terms of RMSE compared with the existing models. The study focuses on different types of methods used to forecast and predict rainfall.

The performance analysis of various kinds of algorithms for rainfall forecasting on weather datasets is done in a study [24]. The main purpose of the work is to develop a multi-scale system for weather forecasting using statistical models and artificial neural networks. The synoptic weights of the artificial neural network were optimised using different types of optimization techniques such as Genetic Algorithm (GA) [25], Hybrid Particle Swarm Optimization and Radial Basis Functions (RBF) [26, 27]. All these techniques were applied to the weather dataset which is collected from different weather stations. A summary of the literature review is shown in Table **1**.

**Table 1. Summary of Literature Review.**

| References | Techniques | Accuracy Measure | Predicting Attributes |
|---|---|---|---|
| In [28]. | MLP, BPN, SVM, RBFN, SOM | RMSE | Min-Max temperature |
| In [29]. | Artificial Neural Network (EBP) | MSE | Temperature |
| In [30]. | Artificial Neural Network (EBP) | RMSE | Min-Max temperature |
| In [31]. | ANN | RMSE | Precipitation, Wind, temperature, |
| In [32]. | Nonlinear Time Series Model | RMSE | Temperature |
| In [33]. | Fuzzy time Series, wheat cops production | RMSE | Linguistic variables |
| In [34]. | ARIMA for rainfall prediction | R2 | Precipitation, temperature, and wind speed |

*(Table 1) cont.....*

| References | Techniques | Accuracy Measure | Predicting Attributes |
|---|---|---|---|
| In [35]. | SARIMA for rainfall forecasting | R2 | Precipitation, temperature, and wind speed |
| In [36]. | Hybrid ARIMA, for rainfall forecasting | MSE | Temperature, precipitation, pressure |
| In [37]. | RNN-LSTM model for weather forecasting | MAE | Average Temperature, Precipitation, wind speed |
| In [38]. | TD-LSTM for temperature forecasting | MAE | Temperature |
| In [39]. | CNN-LSTM for indoor temperature modelling | R2 | Temperature |
| In [40]. | A comparative study of ARMA LSTM, ANFIS-FCM, for prediction of air temperature | R2 | Temperature |
| In [41]. | GANs-LSTM model for soil temperature estimation | R2 | Temperature |

# WEATHER FORECAST USING LINEAR REGRESSION, AUTOREGRESSIVE INTEGRATED MOVING AVERAGE AND LONG-SHORT TERM MEMORY MODEL

This section introduces linear regression; autoregressive integrated moving average and long-short term memory model along with required parameter settings.

## Linear Regression

The term regression was first coined by Francis Galton [42], it is a method to find the statistical relationship between two variables. The application of regression analysis is found in different fields such as agriculture, engineering, economics, and biology. The main purpose of regression analysis is given below:

  i. To establish the causal relationship between the target variable $[x_1, x_2, ... x_n]$.
 ii. To predict target variable y based on a set of values $[x_1, x_2, ... x_n]$.
iii. To find the appropriate regressor variable for target variable y.

The linear regression model is represented in (Equation **2**).

$$y = \beta_0 + \beta_1 x + \varepsilon \tag{2}$$

Here, y is the target variable, $\beta_0$, $\beta_1$, $x$ and $\varepsilon$ represent, y intercept, the slope of the linear regression line, the independent variable and random error, respectively.

The target and independent variables are also known as response and explanatory variables. The general representation of the regression model is represented as (Equation **3**).

$$y = E(y) + \varepsilon \tag{3}$$

Here E(y) is the mathematical expectation of the response variable and it is a linear arrangement of investigative variables $[x_1, x_2, \ldots x_n]$.

The regression is linear if the value of n is equal to one, and if E(y) is a non-linear function, the regression is also said to be non-linear, whereas $Var(\varepsilon) = \sigma^2$ represents the error term. For the experiment of the model with n pairs of data can be denoted in (Equation **4**).

$$y_i = \beta_0 + \beta_1 x_i + \varepsilon_i \quad \text{for i} = 1,2,3, \ldots, n \tag{4}$$

In the linear regression model, slope and intercept are the most important parameters.

In the proposed regression model, the prediction of the maximum temperature of Lucknow is estimated.

**Auto-Regressive Integrated Moving Average (ARIMA)**

The ARIMA model was coined by Box and Jenkins in 1990 [43]. This model is used for different types of forecasting and prediction applications like time series, temperature and sales. The standard notations of this model (p, d, q) replaced the variable with integer values to signal. There are three phases involved in this model as given bellows:

Phase-I: This phase, specifies and identifies a response series and candidates of the model. For that, the model computes auto-correlation, cross-correlations partial, autocorrelations, and inverse autocorrelations by using the IDENTIFY statement.

Phase-II: Performs analytic checks on parameters that have been selected from the first phase, it is performed by an ESTIMATE statement.

Phase-III: The model performs a forecast of the predictive value of the time series variable by using the FORECAST expression which is future values.

The weather dataset can be effectively modeled with the help of the ARIMA model. The authors used this model to forecast average precipitation and analyzed its trends. For weather forecasts, the first phase analyses the static data that contains seasonal trends.

For the hypothesis test, the Dickey fuller test was performed. Critical and p-value help decide whether to accept or reject the hypothesis. If the p-value is less than the significant value, the hypothesis was rejected otherwise it is accepted. For the selected dataset, we performed the experiments only from January to April months only and it showed stationary values. Because that time duration was dry and cool weather in the selected region.

Autocorrelation is an important part of the ARIMA model. The value of autocorrelation indicates how data relates to past values over time. The patterns of long and short terms of the data can be obtained by autocorrelations from different dimensions. A lower value indicates weak relation from the past data [43].

For the experiment's purpose, we used the auto-Arima package available in the Python library. The timestamp difference $Z_t$ can be calculated as given below:

$$Z_t = a_{t+1} - a_t \tag{5}$$

The parameters of the ARIMA (p, d, and q) model are represented as:

P represents the autoregressive terms, d is non-seasonal differences and q is the numbers logged. The dataset is divided into two parts, training and testing before applying to the model. The rolling mean function is used for calculating the moving average. The minimum AIC value indicates the best model. Finally, the RMSE value of the model is used for evaluating its accuracy.

**Long short-term memory (LSTM)**

The principal components of an LSTM cell are depicted in Fig. (2). Here $t$ represents the timestamp, $i_t$ represents the input gate, $f_t$ represents the forget gate, $o_t$ represents the output gate and $f_t$ indicates the candidate value which is arranged as following equations.

$$i_t = \sigma\left(W_{i,x}X_t + W_{i,h}h_{t-1} + b_i\right) \tag{6}$$

$$f_t = \sigma\left(W_{f,x}X_t + W_{f,h}h_{t-1} + b_f\right) \tag{7}$$

$$o_t = \sigma\left(W_{o,x}X_t + W_{o,h}h_{t-1} + b_o\right) \tag{8}$$

$$\tilde{c}_t = tanh(W_{\tilde{c},x}\, X_t + W_{\tilde{c},h}\, h_{t-1} + b_{\tilde{c}}) \tag{9}$$

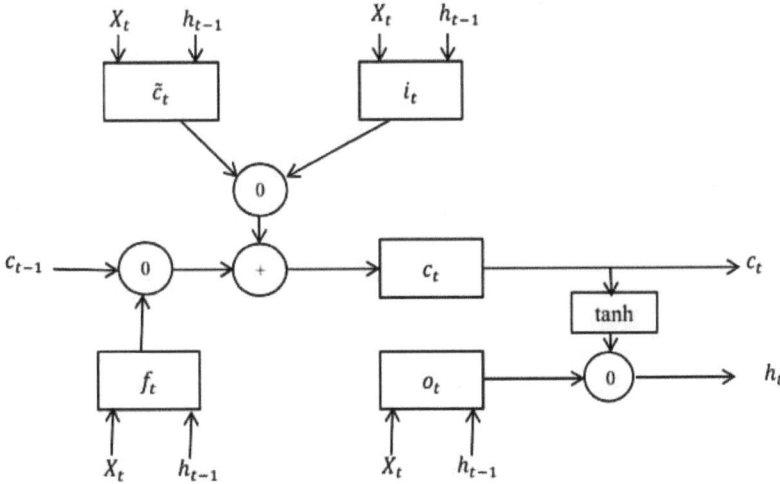

**Fig. (2).** LSTM Cell Structure.

Where $W_{i,x}$, $W_{i,h}$, $W_{f,x}$, $W_{f,h}$, $W_{o,x}$, $W_{c,x}$, and $W_{c,h}$, represent weight matrices, $b_i$, $b_f$, $b_o$ and $b_c$, represent bias vectors, $X_t$, represents current input, $\sigma()$ represents Sigmoid function and $h_{t-1}$ indicates the outcome of an LSTM network at time *t*-1. The role of the forget gate is to calculate the size of the prior memory value that may be clear from the cell state. The input gate identifies new input of cell state which is mathematically formulated as given below:

$$c_t = f_t{}^{\circ} c_{t-1} + i_t{}^{\circ} \tilde{c}_t \tag{10}$$

Here $^{\circ}$ indicates an element-wise product.

The output $h_t$ at time *t* of the network can be calculated as given below:

$$h_t = o_t{}^{\circ} tanh(c_t) \tag{11}$$

Finally, we predict output $y_t$ as given below:

$$\tilde{y}_t = W_y h_t \tag{12}$$

Here, $W_y$ indicates the projection matrix, it is used to minimize the dimension of $h_t$

Fig. (**3**), shows that the feature vector $X_t$ is fed into networks at time $t$, and the cell received feedback $h_{t-1}$ from the previous state. The main objective of training the networks is to minimize the error function $f$ based on the target value $y_t$ as:

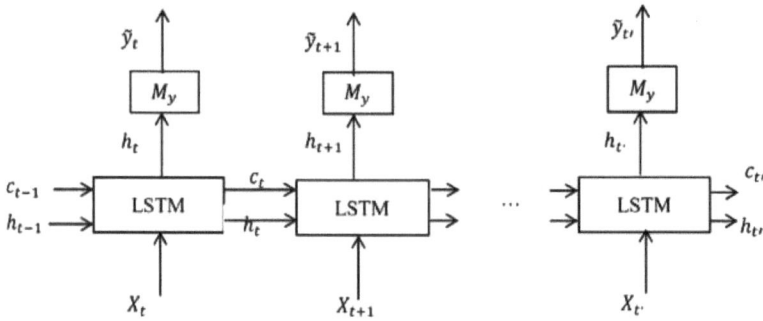

**Fig. (3).** Structure of LSTM network.

$$f = \Sigma_t \; \|y_t - \tilde{y}_t\|^2 \tag{13}$$

by using back-propagation and gradient descent algorithms. In training durations, the values of weights, and bias are auto-adjusted with the help of gradients. Iteration will be completed, when a batch of the dataset is fed into the LSTM network, and learning is done by using a backpropagation algorithm. Due to offline training of the LSTM network, the computation time is less and the network learns very fast.

We performed this experiment using a historical dataset to predict the day-to-day rainfall. In this experiment, the dataset was used from 2019 to 2020. We partition the dataset into training and testing to predict the rainfall.

The Architecture of LSTM networks, as defined in table 2, is implemented by using Keras's deep learning package in python. The input layer of the network had 3 features and 5 timestamps and the values of hidden neurons were 30. The output layer had one neuron with a linear activation function and the value of iteration was set to 40.

**The Architecture of LSTM Network**

model=Sequential ()

model.add(LSTM(30,input_shape=(5,3),return_sequences=True))

```
model.add(Dense(1,activation='linear'))

model.compile(loss='mse',optimizer='adam')

history=model.fit(x_train,t_train,epochs=40,batch_size=50)
```

## EXPERIMENTAL RESULTS

A time-varying weather dataset was collected from the UCI Machine Learning Repository. The dataset contains weather data for Uttar Pradesh state from the year 1999 to 2020. The dataset contains many different types of features such as humidity, temperature, dew point, wind direction, rainfall, *etc*. The authors applied this dataset in different types of machine learning algorithms to forecast weather with different combinations of features.

By implementing the data analysis, we obtain the patterns of humidity, temperature, and rainfall. Figs. (**4-6**), show the patterns of the humidity, temperature, and rainfall in the past years.

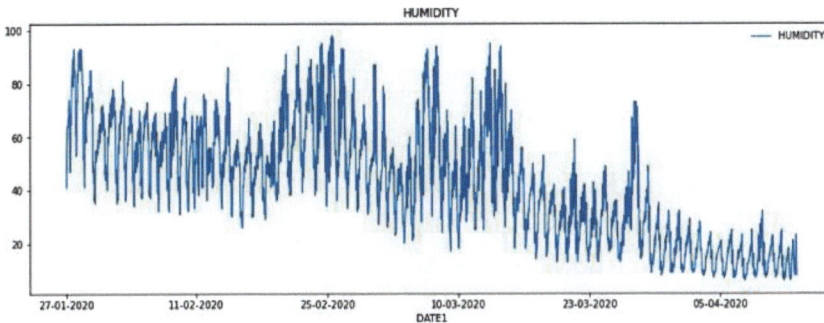

**Fig. (4).** Trends of humidity.

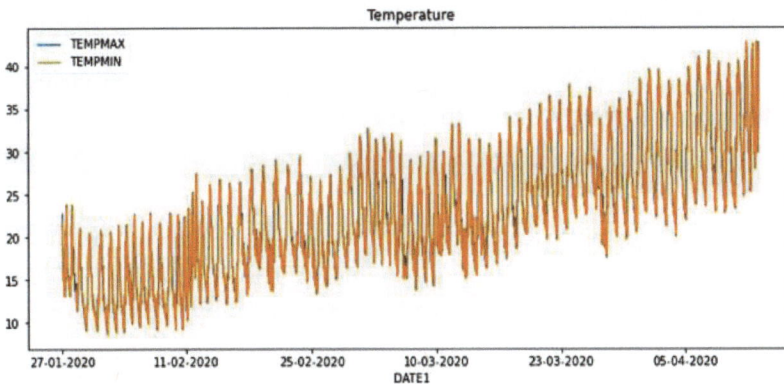

**Fig. (5).** Trends in temperature.

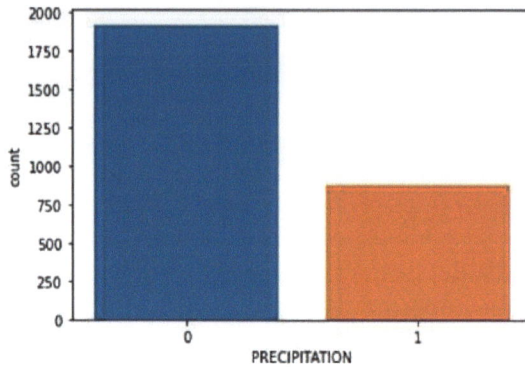

**Fig. (6).** Trends of Rainfall.

The selected data set was applied to the regression algorithm. Here rainfall is an independent variable that is predicted and temperature and humidity are dependent variables. The predicted result is visualized and shown in Fig. (7).

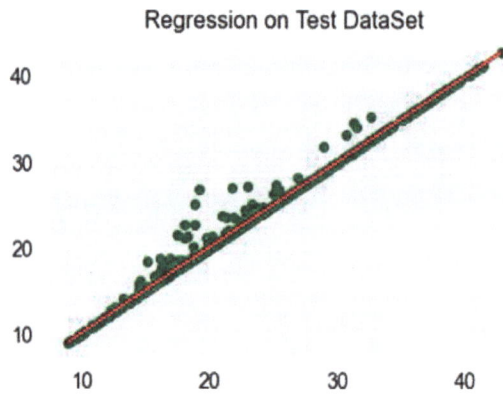

**Fig. (7).** Regression line.

Table **3** shows the accuracy of the model, this is proved by comparing the actual values and the predicted monthly values for the year 2020.

**Table 3. Accuracy of the linear regression**

| Month | Actual value | Predicted value |
|-------|------------|----------------|
| January | 19.30 | 19.413763 |
| February | 20.34 | 20.453675 |
| March | 25.76 | 25.873219 |
| April | 27.82 | 27.933045 |

After the dickey-fuller test, the data was applied to the seasonal decompose function. The autocorrelation function value is 0.73. For the experimental purpose of the Auto-ARIMA model, it is imported from the pmdarima python library [44, 45]. Table **4** shows the AIC values which are useful for selecting the predictors in regression.

**Table 4. Results of AIC Values**

| Model | AIC value |
|---|---|
| ARIMA(0,0,0) | 1088.32 |
| ARIMA(1,0,0) | 902.015 |
| ARIMA(1,0,2) | 845.325 |
| ARIMA(2,0,3) | 900.232 |
| ARIMA(2,0,1) | 902.154 |
| ARIMA(0,0,3) | 920.134 |

Table **4** shows different models, here the minimum AIC value model is selected because it denotes the log value set between 1, 0. Finally, the optimal parameters were selected to fit the model to determine the coefficients of the regression. The forecast model is shown in Fig. ( **8**).

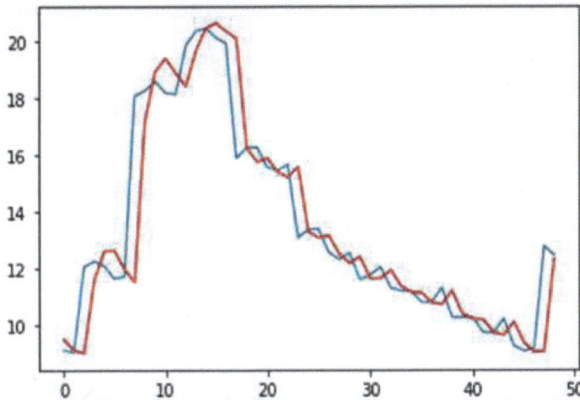

**Fig. (8).** Temperature forecast with ARIMA Model.

The outcome of the state-of-the-art is especially dependent on the collected data of a particular area. The RMSE value of the model is 2.56. However, in the LSTM model shown in Fig. (**9**), 1.5 RMSE value for train data and 1.51 for test data which is better than the ARIMA model.

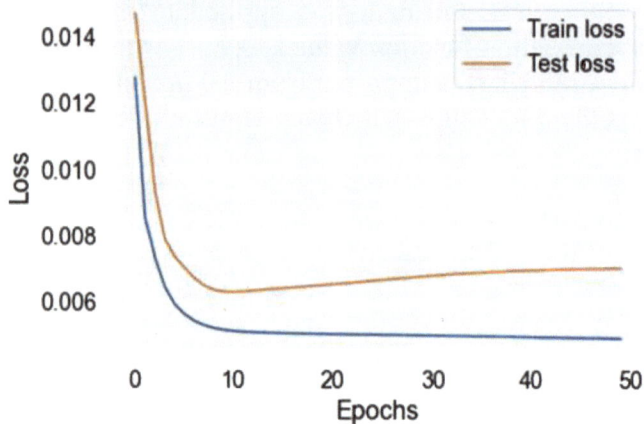

**Fig. (9).** RMSE value for LSTM Model.

Based on the above experiments, it has been observed that LSTM performs better prediction than the ARIMA model when the nature of the dataset is non-linear. In the case of linear datasets, linear regression algorithms performed well.

## CONCLUSION

Weather forecasting has been a challenging field for research because of the time-varying nature of data. It is a broad area of research, there are numerous algorithms proposed to predict temperature, irradiance, maximum precipitation, humidity, and rainfall forecasting on weather datasets with remarkable accuracy. The continuous data can enhance the results of the LSTM model rather than linear regression and the ARIMA model. This study shows that the LSTM and ARIMA model predicts the hourly temperature for a particular region and year. The future scope of this study is to forecast different environmental parameters with improved accuracy.

## REFERENCES

[1]     W.S. Noble, "What is a support vector machine?", *Nat. Biotechnol.,* vol. 24, no. 12, pp. 1565-1567, 2006.
        [http://dx.doi.org/10.1038/nbt1206-1565] [PMID: 17160063]

[2]     M.W. Gardner, and S.R. Dorling, "Artificial neural networks (the multilayer perceptron)—a review of applications in the atmospheric sciences", *Atmos. Environ.,* vol. 32, no. 14-15, pp. 2627-2636, 1998.
        [http://dx.doi.org/10.1016/S1352-2310(97)00447-0]

[3]     B. Leo, "Random forests", *Machine learning,* vol. 21, no. 1, pp. 90-117, 2001.

[4]     S. Agatonovic-Kustrin, and R. Beresford, "Basic concepts of artificial neural network (ANN) modeling and its application in pharmaceutical research", *J. Pharm. Biomed. Anal.,* vol. 22, no. 5, pp. 717-727, 2000.
        [http://dx.doi.org/10.1016/S0731-7085(99)00272-1] [PMID: 10815714]

[5]     W.S. McCulloch, and W. Pitts, "A logical calculus of the ideas immanent in nervous activity", *Bull. Math. Biophys.,* vol. 5, no. 4, pp. 115-133, 1943.
[http://dx.doi.org/10.1007/BF02478259]

[6]     H. Leung, and S. Haykin, "The complex backpropagation algorithm", *IEEE Trans. Signal Process.,* vol. 39, no. 9, pp. 2101-2104, 1991.
[http://dx.doi.org/10.1109/78.134446]

[7]     R. Caruana, and A. Niculescu-Mizil, "An empirical comparison of supervised learning algorithms", *Proceedings of the 23rd international conference on Machine learning,* 2006pp. 161-168
[http://dx.doi.org/10.1145/1143844.1143865]

[8]     K. Cho, B. Van Merriënboer, C. Gulcehre, D. Bahdanau, F. Bougares, H. Schwenk, and Y. Bengio, "Learning phrase representations using RNN encoder-decoder for statistical machine translation", *arXiv,* 2014.
[http://dx.doi.org/10.3115/v1/D14-1179]

[9]     A. Sherstinsky, "Fundamentals of recurrent neural network (RNN) and long short-term memory (LSTM) network", *Physica D,* vol. 404, p. 132306, 2020.
[http://dx.doi.org/10.1016/j.physd.2019.132306]

[10]    E.I. Vlahogianni, J.C. Golias, and M.G. Karlaftis, "Short-term traffic forecasting: Overview of objectives and methods", *Transp. Rev.,* vol. 24, no. 5, pp. 533-557, 2004.
[http://dx.doi.org/10.1080/0144164042000195072]

[11]    "Electricity load and price forecasting webinar case study". Available From: http://www.math works.com/matlabcentral/fileexchange/28684-electricity-load-and-price-forecasting-webin-r-case-study

[12]    M. Diagne, M. David, P. Lauret, J. Boland, and N. Schmutz, "Review of solar irradiance forecasting methods and a proposition for small-scale insular grids", *Renew. Sustain. Energy Rev.,* vol. 27, pp. 65-76, 2013.
[http://dx.doi.org/10.1016/j.rser.2013.06.042]

[13]    M. Ali, R. Prasad, Y. Xiang, and Z.M. Yaseen, "Complete ensemble empirical mode decomposition hybridized with random forest and kernel ridge regression model for monthly rainfall forecasts", *J. Hydrol. (Amst.),* vol. 584, p. 124647, 2020.
[http://dx.doi.org/10.1016/j.jhydrol.2020.124647]

[14]    Y. Sharma, and S. Sisodia, "Temperature prediction based on Swain and MTPSO with automatic clustering algorithm", *2nd International Symposium on Computational and Business Intelligence,* 2014pp. 101-105

[15]    N.N. Arjun, V. Prema, D. Kumar, P. Prashanth, V. Preekshit, and U. Rao, "Multivariate regression models for prediction of wind speed", *International Conference on Data Science and Engineering, ICDSE,* 2014pp. 171-176
[http://dx.doi.org/10.1109/ICDSE.2014.6974632]

[16]    S. Kavitha, J. Rajesh Banu, J. Vinoth Kumar, and M. Rajkumar, "Improving the biogas production performance of municipal waste activated sludge via disperser induced microwave disintegration", *Bioresour. Technol.,* vol. 217, pp. 21-27, 2016.
[http://dx.doi.org/10.1016/j.biortech.2016.02.034] [PMID: 26897472]

[17]    W. Shen, W. Xiao, and X. Wang, "Passenger satisfaction evaluation model for Urban rail transit: A structural equation modeling based on partial least squares", *Transp. Policy,* vol. 46, pp. 20-31, 2016.
[http://dx.doi.org/10.1016/j.tranpol.2015.10.006]

[18]    A. Mathew, S. Khandelwal, and N. Kaul, "Investigating spatial and seasonal variations of urban heat island effect over Jaipur city and its relationship with vegetation, urbanization and elevation parameters", *Sustain Cities Soc.,* vol. 35, pp. 157-177, 2017.
[http://dx.doi.org/10.1016/j.scs.2017.07.013]

[19]   M. Pushpa, and K.K. Patil, "Deep learning based weighted SOM to forecast weather and crop prediction for agriculture application", *Int. J. Intell. Eng. Syst,* vol. 11, pp. 167-176, 2018.

[20]   C. Lennard, and G. Hegerl, "Relating changes in synoptic circulation to the surface rainfall response using self-organising maps", *Clim. Dyn.,* vol. 44, no. 3-4, pp. 861-879, 2015.
[http://dx.doi.org/10.1007/s00382-014-2169-6]

[21]   A. Geetha, and G.M. Nasira, "Data mining for meteorological applications: Decision trees for modeling rainfall prediction", *IEEE international conference on computational intelligence and computing research.,* 2014pp. 1-4
[http://dx.doi.org/10.1109/ICCIC.2014.7238481]

[22]   Q. Luan, X. Fang, C. Ye, and Y. Liu, "An integrated service system for agricultural drought monitoring and forecasting and irrigation amount forecasting", *23rd International Conference on Geoinformatics,* 2015pp. 1-7
[http://dx.doi.org/10.1109/GEOINFORMATICS.2015.7378617]

[23]   B. Jabber, K. Rajesh, D. Haritha, C.Z. Basha, and S.N. Parveen, "Rainfall prediction using machine learning & deep learning techniques", *2020 International Conference on Electronics and Sustainable Communication Systems (ICESC),* 2020pp. 92-97

[24]   L. Naveen, and H.S. Mohan, "Atmospheric weather prediction using various machine learning techniques: A survey", *3rd International Conference on Computing Methodologies and Communication (ICCMC),* 2019pp. 422-428

[25]   D.S. Weile, and E. Michielssen, "Genetic algorithm optimization applied to electromagnetics: A review", *IEEE Trans. Antenn. Propag.,* vol. 45, no. 3, pp. 343-353, 1997.
[http://dx.doi.org/10.1109/8.558650]

[26]   J. Wu, J. Long, and M. Liu, "Evolving RBF neural networks for rainfall prediction using hybrid particle swarm optimization and genetic algorithm", *Neurocomputing,* vol. 148, pp. 136-142, 2015.
[http://dx.doi.org/10.1016/j.neucom.2012.10.043]

[27]   J.M. Stanton, "A brief history of linear regression for statistics instructors", *J. Stat. Educ.,* vol. 9, no. 3, 2001.

[28]   D.R. Nayak, A. Mahapatra, and P. Mishra, "A survey on rainfall prediction using artificial neural network", *Int. J. Comput. Appl.,* vol. 72, no. 16, 2013.

[29]   D.B. Shank, G. Hoogenboom, and R.W. McClendon, "Dewpoint temperature prediction using artificial neural networks", *J. Appl. Meteorol. Climatol.,* vol. 47, no. 6, pp. 1757-1769, 2008.
[http://dx.doi.org/10.1175/2007JAMC1693.1]

[30]   C. Robinson, and N. Mort, "A neural network system for the protection of citrus crops from frost damage", *Comput. Electron. Agric.,* vol. 16, no. 3, pp. 177-187, 1997.
[http://dx.doi.org/10.1016/S0168-1699(96)00037-3]

[31]   M.P. Darji, V.K. Dabhi, and H.B. Prajapati, "Rainfall forecasting using neural network: A survey", *2015 international conference on advances in computer engineering and applications.,* 2015pp. 706-713
[http://dx.doi.org/10.1109/ICACEA.2015.7164782]

[32]   W. Wu, Y. Xiao, G. Li, W. Zeng, H. Lin, S. Rutherford, Y. Xu, Y. Luo, X. Xu, C. Chu, and W. Ma, "Temperature–mortality relationship in four subtropical Chinese cities: A time-series study using a distributed lag non-linear model", *Sci. Total Environ.,* vol. 449, pp. 355-362, 2013.
[http://dx.doi.org/10.1016/j.scitotenv.2013.01.090] [PMID: 23454696]

[33]   S.R. Singh, "A simple method of forecasting based on fuzzy time series", *Appl. Math. Comput.,* vol. 186, no. 1, pp. 330-339, 2007.
[http://dx.doi.org/10.1016/j.amc.2006.07.128]

[34] S. Swain, S. Nandi, and P. Patel, Development of an ARIMA model for monthly rainfall forecasting over Khordha district, Odisha, India.*Recent Findings in Intelligent Computing Techniques.* Springer: Singapore, 2018, pp. 325-331.
[http://dx.doi.org/10.1007/978-981-10-8636-6_34]

[35] M. Nirmala, and S.M. Sundaram, "A seasonal ARIMA model for forecasting monthly rainfall in Tamilnadu", *National Journal on Advances in Building Sciences and Mechanics,* vol. 1, no. 2, pp. 43-47, 2010.

[36] P. Unnikrishnan, and V. Jothiprakash, "Hybrid SSA-ARIMA-ANN model for forecasting daily rainfall", In: *Water Resources Management* vol. 34. , 2020, no. 11, pp. 3609-3623.
[http://dx.doi.org/10.1007/s11269-020-02638-w]

[37] S. Mittal, and O.P. Sangwan, "Big data analytics using deep LSTM networks: A case study for weather prediction", *Adv. Sci. Technol. Eng. Syst.,* pp. 133-137, 2020.

[38] J. Liu, T. Zhang, G. Han, and Y. Gou, "TD-LSTM: Temporal dependence-based LSTM networks for marine temperature prediction", *Sensors,* vol. 18, no. 11, p. 3797, 2018.
[http://dx.doi.org/10.3390/s18113797] [PMID: 30404217]

[39] F. Elmaz, R. Eyckerman, W. Casteels, S. Latré, and P. Hellinckx, "CNN-LSTM architecture for predictive indoor temperature modeling", *Build. Environ.,* vol. 206, p. 108327, 2021.
[http://dx.doi.org/10.1016/j.buildenv.2021.108327]

[40] A. Ozbek, A. Sekertekin, M. Bilgili, and N. Arslan, "Prediction of 10-min, hourly, and daily atmospheric air temperature: Comparison of LSTM, ANFIS-FCM, and ARMA", *Arab. J. Geosci.,* vol. 14, no. 7, p. 622, 2021.
[http://dx.doi.org/10.1007/s12517-021-06982-y]

[41] Q. Li, H. Hao, Y. Zhao, Q. Geng, G. Liu, Y. Zhang, and F. Yu, "GANs-LSTM model for soil temperature estimation from meteorological: A new approach", *IEEE Access,* vol. 8, pp. 59427-59443, 2020.
[http://dx.doi.org/10.1109/ACCESS.2020.2982996]

[42] S. Senn, "Francis Galton and regression to the mean", *Significance,* vol. 8, no. 3, pp. 124-126, 2011.
[http://dx.doi.org/10.1111/j.1740-9713.2011.00509.x]

[43] S.C. Hillmer, and G.C. Tiao, "An ARIMA-model-based approach to seasonal adjustment", *J. Am. Stat. Assoc.,* vol. 77, no. 377, pp. 63-70, 1982.
[http://dx.doi.org/10.1080/01621459.1982.10477767]

[44] K. Kalpakis, D. Gada, and V. Puttagunta, "Distance measures for effective clustering of ARIMA time-series", *Proceedings IEEE international conference on data mining,* 2001pp. 273-280
[http://dx.doi.org/10.1109/ICDM.2001.989529]

[45] T.W. Yoo, and I.S. Oh, "Time series forecasting of agricultural products' sales volumes based on seasonal long short-term memory", *Applied Sciences,* vol. 10, no. 22, p. 8169, 2020.
[http://dx.doi.org/10.3390/app10228169]

## CHAPTER 7

# Intelligent Crop Planning and Precision Farming

**Vani Agrawal**[1,*]

[1] *Department of Computer Science and Application, School of Engineering and Technology, ITM University, Gwalior, India*

**Abstract:** Countries are more concerned about agricultural needs as it is considered to be the essential source of one's life. In our country, agriculture and farming play a vital role in the economy and provide 45% of overall support for economic development. In earlier research, the development of various agricultural support devices was introduced but all were stuck up to a certain level. Remote surveillance, SMS-based agricultural watering systems, and management are a few implementations in the area. But all these implementations are facing some technical challenges due to their complexity and it is hard to maintain the accuracy of these systems. Today's agricultural demands can be supported by Precision agriculture and Intelligent Crop Farming. This chapter focuses on different aspects of Precision Agriculture and Smart Farming.

**Keywords:** Artificial intelligence, Climate smart agriculture, Intelligent crop planning, Precision farming.

## INTRODUCTION

The word "agriculture" is derived from two Latin words, 'agricultura' from "ager" *i.e.* "field" and "cultura" *i.e.* "cultivation" or "growing". In common sense, agriculture is typically described as farming. It is an art and science of prudent endeavors to reshape a part of the earth's crust through the cultivation of plants and other crops as well as raising livestock for sustenance or other necessities for human beings and economic gain.

In layman's definition, farming is the basic process of food production which includes the in-depth use of traditional knowledge, land, known tools, natural resources, fertilizers, and the cultural faith of the farming society. But a production system solely based on traditional methods and cultural beliefs is inevitably followed by superstitions and speculative judgment in disguise of skilled predictions. Such   an approach through repeated failures can be useful at

---

[*] **Corresponding author Vani Agrawal:** Department of Computer Science and Application, School of Engineering and Technology, ITM University, Gwalior, India; E-mail: vaniagrawal.mca@gmail.com

one point in time after an innumerable number of experiments and resources are lost, it still needs for an upgrade that can reduce the problems and help farmers make informed decisions instead of speculative ones. And here is where modern technology comes into play. With the advent of Information Technology, nowadays scientists have come up with methods and technologies that are data-driven and take into account factors contributing to success and failures alike calculating the best possible route; thus recommending the farmers the precise action that needs to be taken to reach a specific end goal. This method of farming based on data intelligence is what we are calling precision farming.

To understand precision farming, we must understand how precise we can be by using the technology at our disposal; and here comes our use of data science and analytics. Data science is a field that deals with large volumes of data and its study uses modern tools and techniques to look for unknown patterns and derive from them meaningful information which can help in taking business decisions. Data science uses complex machine learning algorithms to build predictive models. The data used for analysis can come from many different sources and be presented in various formats. Now let us consider the attributes in farming such as plants, plant type, species, weather and climate, soil, *etc.* assets of data. Data Science algorithms and techniques will tell us precisely, when and at what time, what species of plant will be the most suited to the condition of soil structure at the moment and what weather to expect, and what steps are needed to be taken for proper maintenance concerning the weather and previous performances of other farmers on the same piece of land. This gives an exact farming model which will help in planting intelligently with precise actions that we will be taking for minimizing the loss of resources, time and overall expenditure and also giving the best yield and taking the current yield as another attribute. All of this coupled with the latest trends in agricultural technology gives us the ability to do precision planning which in turn helps us achieve an intelligent approach to farming.

## Precision Farming

The precision farming approach provides the precise amount of required inputs to achieve increased production yields in comparison with traditional cultivation techniques. It is the system through which information, technology, and management can be employed, to increase production efficiency, improve product quality, improve the efficiency of crop chemical use, conserve energy and protect the environment. Fig. (**1**) depicts the steps for precision farming.

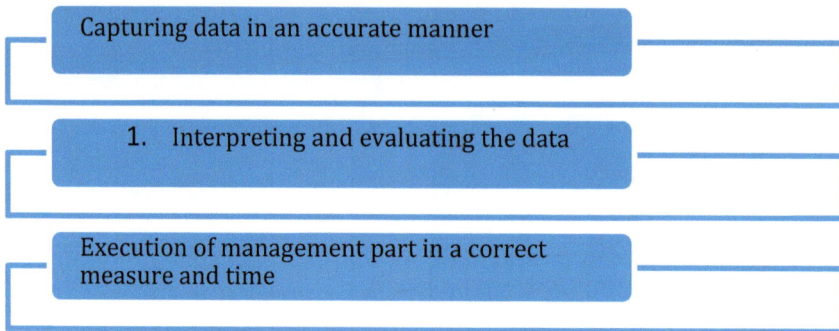

**Fig. (1).** Steps for precision farming.

Precision farming is a trending idea that naturally leads the way for higher profitability and fewer ill effects on the environment. Precision agricultural breakthroughs of today may provide the technology for the environmentally-friendly agriculture of tomorrow. Precision farming promises large yield improvement with low external input use. It is exponentially beneficial for small farmers in developing nations.

Multiple information to farmers from Precision agriculture can be listed as [1]-having better knowledge of the plant needs:

1. Real-time farm information

2. Helps the farmer in prompt decision making

3. Easy to trace

4. Enhance farm products quality

**Need for Precision Farming**

Traditional Agriculture faces educational and economic challenges as per the research. Aspects like the lack of local experts, funds, knowledgeable research, and extension personnel also add to these existing challenges in education [2]. There is a major need for precision agriculture in India to increase crop productivity and optimal use of water resources. The unnecessary use of heavy chemicals proves hazardous in crop production, which can be reduced by precision agriculture. Additionally, it helps in the prevention of soil degradation. Dissemination of modern farm practices to improve the quality, quantity, and reduced cost of production can be done through it. This way, this modern approach can also help in improving the socioeconomic status of farmers.

Precision farming can be understood as innovative agriculture which can achieve high productivity of crops as shown in Fig. (**2**). It is based on the identification of spatial and temporal variability in crop production. To aim for maximizing productivity and lowering environmental risks, it becomes important to take variability into account.

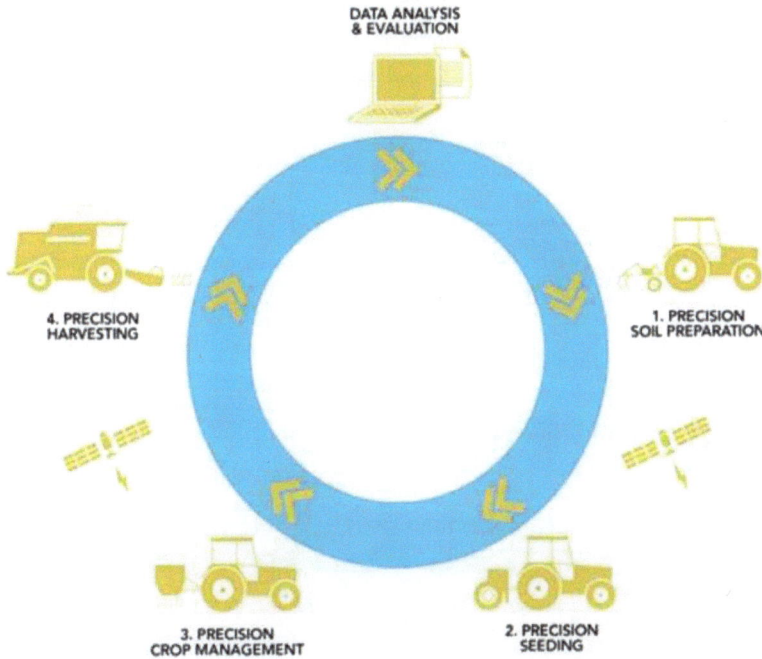

**Fig. (2).** Precision Farming Simplified.
(Source: https://www.scenario.co.za/en-ZA/News/Article/View/basic-concepts-of- precision-farming)

## PRECISION FARMING AND CHANGING TIMES

### Past

The beginning of the 80s witnessed the use of precision farming in the form of yield mapping in the US where corn and soybean were cultivated using crop rotation [3, 4]. Massey Ferguson MF 38 (A combine harvester) having the technology for yield mapping was utilized by John P. Fenton for crop rotation involving common wheat and bean later in 1988 and from 1988 to 1993, the yield mapping and study of soil parameters like P, K, Mn were carried out [5]. The year 1997 saw combined harvesters with mounting technology getting commercialized in Europe [6]. Italy saw the usage of precision farming when crops were cultivated at the end of 1990 and later to check the efficiency of yield mapping, it was applied in 3 zones of the same field (deep, minimum, and no-tillage) [7].

Hence, the past observed precision farming in the form of-

1. Yield mapping

2. Fertilizer application at a spatially variable rate

Tree crops were exposed to yield mapping for the first time in the USA in 1998 only. Grape yield sensors were commercialized in some regions of South Australia thus the emergence of precision viticulture [8 - 12]. Vegetative vigor mapping technology, yield mapping technology and some spatially variable field operations [13] were some ways used for grapevine precision.

## Present

The present is witnessing the use of precision agriculture in industrialized countries alongside large low-profit cereal farms and highly profitable grapevine farms. It is well known in countries like the USA, Germany, Denmark, Netherlands, and UK [14].

Real-time forms of precision are emerging for weed control in North America for herbaceous crops [15].

Therefore, applications can be categorized into two sections: [16]

1. For herbaceous crop

• Yield map technology

• Application of herbicide and fertilizer at a spatially variable rate

2. For grapevine

• Field operations at a spatially variable rate

• Yield maps

• Vigour mapping

Additionally, for accessing soil health, on-the-go mapping sensors are also commercially available [17].

## Precision Farming: Scenario of India

The idea of precision farming is still in its infancy in India which means that there exists a very large group of farmers who do not know that there lies a concept like precision farming at all. Companies like Tata Chemicals Limited and Trimble are leading the way for farmers to make them aware of the concept to assess soil conditions, examining crop health, pest invasions, and crop yield prediction through remote sensing. They have taken the lead to guide farmers for surplus crop production thus earning more profits.

A recent step in the direction was taken by ICAR-IARI in the form of a session on "Sensors and Sensing of Precision Agriculture".

Precision agriculture means 'right-input' at the 'right-time' in the 'right-amount' at the 'right place' and in the 'right-manner' for improving productivity, conserving natural resources, and avoiding any ecological or social tribulations, as per the Indian Journal of Fertilization [18].

**Precision Farming: An add on**

The increasing population demands an increase in agricultural production to feed the growing population and to keep pace with this demand, and innovation to the traditional methods is a dire need that comes in the form of precision agriculture. The only way ahead is to grow with the technology and utilize it for the fullest gain. Precision agriculture can help as shown in Fig. (**3**).

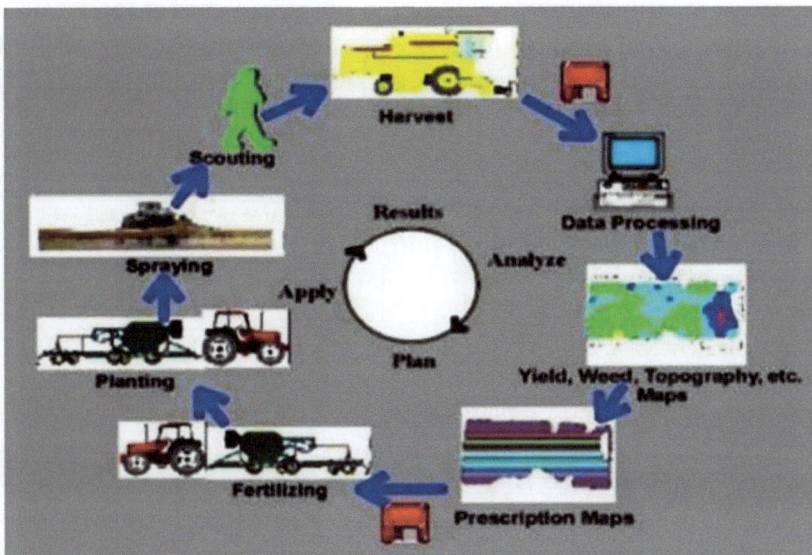

**Fig. (3).** Precision Agriculture Cycle.

1. Reduces unrequired chemical usage for crop production.

2. Preventing soil degradation thus meeting the goal of sustainable agriculture.

3. Easy surveillance of agricultural fields.

4. Improving the quality and quantity of crops by increasing the chances of the real-time availability of required information.

5. Addressing the requirements of individual fields.

6. Putting effort into environmental protection through the use of chemicals, only when required.

7. Managing the resources in a better way thus reducing their wastage.

(Source: https://www.iotforall.com/smart-farming-future-of-agriculture)

**Tools and Techniques Used for Precision Farming**

*Global Positioning System (GPS)*

Satellites of GPS broadcast signals in a way to alert the GPS receivers about the location. The information is received in real-time, which means that time-to-time updates on position are received while the satellites are in motion. The updates about the precise location allow soil and crop measurements to reflect on the map. In the field, these GPS receivers are carried or mounted allowing users to reach specific locations for collecting samples and also aid in the treatment of affected areas.

The US Department of Defense (DoD) was the responsible authority to maintain NAVSTAR GPS. Although designed as a military system but it was made available to civilians for positioning, putting some restrictions. It was carefully designed with a full set of 24 satellites that reached full operational capability [30].

The information about field boundaries mapping, roads, irrigation systems, and problematic crop areas such as weeds or diseases is provided by receivers with the help of GPS technology. The accuracy of information helps farmers to create farm maps with precise acreage for field areas, road locations, and distances between points of interest. It directs farmers to accurately navigate to specific locations in the field to collect soil samples or monitor crop conditions [19].

The use of GPS in agriculture (Fig. **4**) is emerging and has proved to be a game-changer. This has become a vital step in guiding farmers to follow a data-driven approach to agriculture, increase production, and address food security [20].

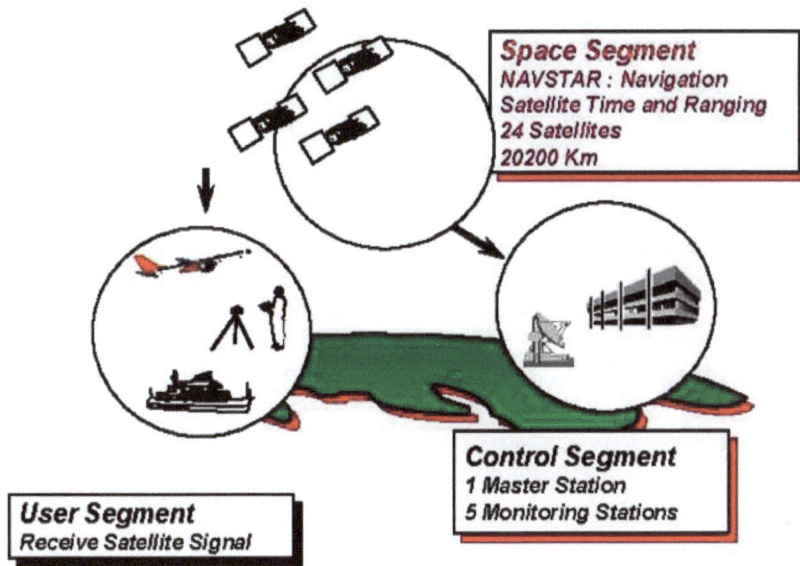

**Fig. (4).**   Components of GPS. (Source: https://slidetodoc.com/gps-basic-theory-contents-gps-general-characteristics-gps/)

GPS allows users to record positional data (in terms of latitude, longitude, and elevation) with a precision of 100 to 0.01 meters. It also provides information related to soil category, pest amount, weed incursion, water holes, boundaries, and obstructions. A light/sound guidance panel (DGPS), antenna, and receiver form the autonomous control system. Farmers can use the system to correctly locate field locations to make sure the effective application of inputs (seeds, fertilizers, pesticides, herbicides, and irrigation water).

**Tractor Guidance:** Initially, there were no technologies to put the tractors in auto-pilot mode. But if a GPS system is installed in the tractor then the steps to follow can be programmed – for cultivation, fertilization, pest control, and harvest. This programming can help in saving a large chunk of money.

**Cropduster Targeting:** The attacks of insects/pests are not uniform, they prefer some areas more than others to attack. These attack-prone areas can be tracked through GPS and cropduster pilots can utilize this recorded data to target the specific areas instead of treating the entire field which can in turn help in saving time and resources and the extent of crop exposure to insecticide can be reduced.

**Tracking Livestock:** GPS transmitters can be used to monitor the location of valuable animals on a large farm.

**Yield Monitoring:** Fields are divided into zones for the estimation of yield variations across the property and each zone is estimated and plotted on a map. This map can help in understanding the property and in making decisions about the next planting.

**Soil Sampling:** Mapping software and GPS can help in the collection of soil samples across a large property. The data received from the laboratory can be plotted on the maps and decisions for soil treatment can be made for various parts of the property. This saves money and time only by treating the marked areas as per the need.

### Sensor Technologies

Electromagnetic conductivity, photo electricity, ultrasound, *etc.* technologies are used to calculate humidity, vegetation, temperature, texture, structure, nutrient level, *etc.*

Sensors allow the easy collection of large quantities of data without laboratory analysis. Remote sensing data are utilized to differentiate crop species, locate stress conditions, identify pests and weeds, and monitor drought, soil, and plant conditions.

LM35 and UFM-M1 sensors were utilized for precision farming to measure the aspects like temperature and pressure respectively. Another technique used for precision farming is the development of an algorithm that forms the basis of network hierarchy and communication model. The concept of MFC (Microbial Fuel cell) which can convert soil energy into electrical energy through microbes and with the help of some acetates, which are an efficient energy resource for supplying power to the sensors, has also been used to study precision agriculture [21].

### Geographic Information System (GIS)

GIS hardware installed with software links the available information of different places in one place, which can be explored as per requirement, and then compilation, storage, retrieval, and analysis of the feature attributes and location data help in making maps. It helps in managing spatial information and farmers in decision-making. Information about soil type, crop yield, quality, and field boundaries can be collected through GIS.

Usually, the experience of farmers helps them distinguish between the yielding areas in terms of the high and low yield areas of the farm and helps them identify the areas that are more prone to problems [22].

## Grid Soil Sampling and Variable-rate Fertilizer (VRT) Application

Numerous farming operations can be done with the help of variable-rate technologies (VRT) which have automatic operations that can be used with or without a GPS [23]. VRT helps the farmers which allows for a variable input program so that farmers can adjust the number of inputs applied in a specific location. This technology manages the number of seeds, fertilizers, insecticides, and water for the farmers and also optimizes planting density and increases the effectiveness of pesticide and nutrient application rates, lowering farm costs and, most significantly, lowering negative environmental impact. VRT technology enhances input efficiency, field profitability, and environmental responsibility [24]. The key components of this technology are a system, programs, a sensor, and a differential global positioning system (DGPS) [25].

## Crop Management

Understanding many aspects like the variation in soil conditions and topography which influence crop performance within the field is possible due to the received satellite data. It also makes possible the management of seeds, fertilizers, pesticides, herbicides, and water control, precisely to increase the yield.

## Soil and Plant Sensors

Information on soil properties and plant fertility/water status can be derived from these sensors. Electrical conductivity (EC) sensors are used to describe soil variability while surveying the field and also provide real-time information when used over the field. The EC is sensitive to changes in the soil texture and salinity, thus it provides an exact way to implement site-specific management.

## Rate Controllers

Rate controllers are used to control the rate by which chemical inputs like fertilizers and pesticides should be delivered. These help in monitoring the velocity rate of the tractor/sprayer traveling across the field, along with the flow rate and pressure of the material, and making the required adjustments in real-time.

## Precision Irrigation in Pressurized Systems

Sprinkler irrigation controlling the motion of irrigation machines with GPS-based controllers is the trending development. In order to monitor soil and other ambient conditions, wireless communication, and sensor technologies are being developed along with operation parameters of the irrigation machines to target higher and more efficient water applications.

## *Software*

With the help of the software, the display-controller interfacing, information layers mapping, pre and post-processing data analysis and interpretation, farm accounting of inputs per field, *etc.* can be possible. The most common work of software:

- Map generation (*e.g.* yield, soil)
- Filtration of data collected ;
- Variable rate applications maps generation (*e.g.* for fertilizer, lime, chemicals);
- Overlaying of different maps;
- Help with advanced geostatistical features.

## Intelligent Crop Planning

Intelligent crop planning or smart farming or Intelligent farming is simply the concept of growing more with less effort. It is all about implementing data in such a way that the labor of humans is utilized effectively and the crop can be enriched, qualitatively and quantitatively.

Traditional methods of farming are deeply infused in the practices of farming which can't be parted easily but with changing requirements, there lies a lot of scope for improvement in order to achieve the aim to fulfill the enhanced needs. Smart farming, therefore, comes as a solution that comes with informed and data-driven decisions and these decisions can also cater to the problem of the excessive workforce which can utilize its potential to perform some value-added activities.

It is a little different from precision farming in the sense that here the focus stays on capturing data and interpreting that through computational technologies as shown in Fig. (**5**), thus making that more useful and predictable. Following are used as a tool for smart farming [26]:

- Tablets/Phones
- Artificial Intelligence
- Internet of Things
- Unmanned Aircraft System
- Robotics
- Sensors
- Drones

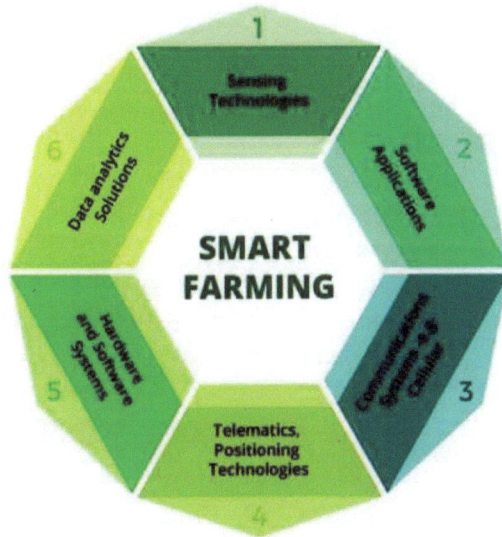

**Fig. (5).**  Tools for Smart Farming. (Source: https://www.iotforall.com/smart-farming-future-of-agriculture) (Source: https://www.wipro.com/holmes/towards-future-farming-how-artificial-intelligence-is-transforming-the-agriculture-industry)

## Intelligent Crop Planning and Artificial Intelligence

The emergence of Artificial Intelligence pushed traditional Indian Agriculture for a big revolution. The major role of AI comes in the form of increased productivity. AI can also address the problem of resource scarcity be it materials or humans in the field of farming. It can also help in dealing with multiple complexities. The role of AI can be summarized as shown in Fig. (**6**).

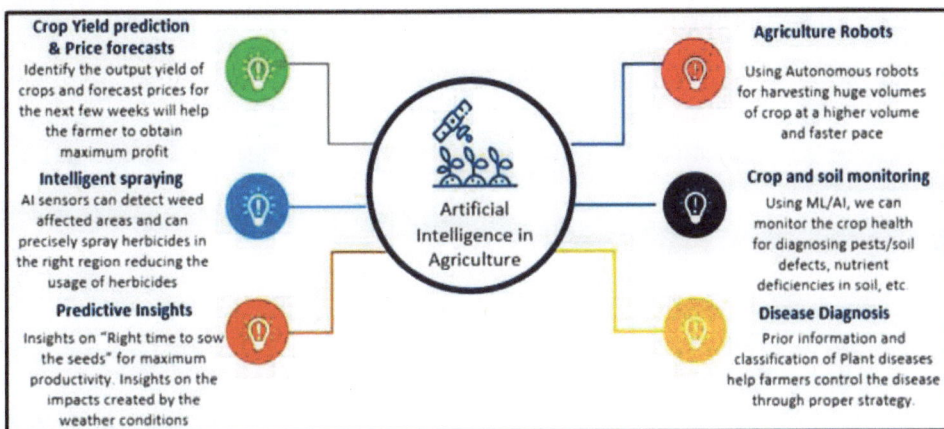

**Fig. (6).**  Role of AI in Agriculture.

## Climate-smart Agriculture

With the rising levels of environmental concerns, there is a dire need of developing agriculture that is climate-smart. In addition to the challenge of food security, the challenge of the increasing vulnerability of agriculture to climate also exists which demands Climate-smart agriculture. The challenges like rising in temperature, variability in weather, invasive species attacks, and a shift in agroecosystem boundaries are proving deadly for crop production. These reduce the crop yield, degrade the nutritional quality and also lower the productivity of livestock and this needs to be addressed to meet the rising demand. For this, the concept of climate-smart agriculture stands relevant which is an all-inclusive approach as depicted in Fig. (**7**). It aims at managing landscapes plus meeting the challenges of food security along with keeping pace with changing climate.

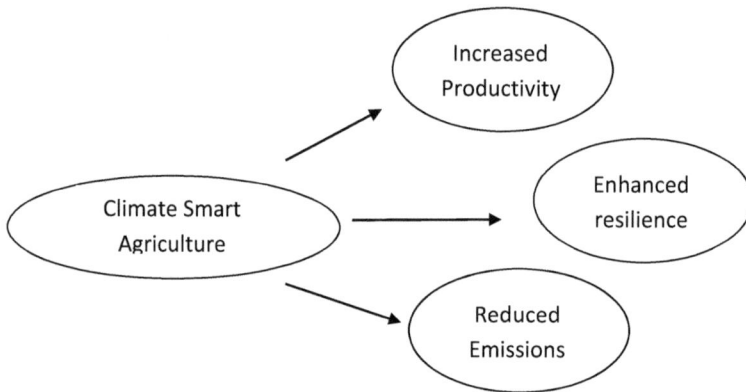

**Fig. (7).** Importance of Climate-Smart Agriculture.

## Challenges that Remain

Precision farming and Intelligent Crop Planning are the leading ways to adapt to the changing scenario of food and production demand. However, there is more than one aspect that creates the hurdle in achieving them fully and these can be as under:

## Data

The advanced solutions in the form of Precision Farming and Intelligent Crop planning put us with the high expectation of desired outcomes. Target input, fewer fertilizers, and other environment-friendly aspects make these more beneficial [27, 28] but there are times when the intensity of these effects is not really known [28]. The data to study is not adequately available and there stays a lot of scope to study the observed impacts.

Sometimes the data which is not properly understood is what the market is flooded with. The raw data in the form of high-resolution images, yield maps, data points, *etc.* come every now and then which confuses the farmers.

## Infrastructure

The small size of landholdings limits the success of precision farming and intelligent crop planning. Poor infrastructure proves to be a big problem. Even the infrastructure to develop and test the technologies is not adequately available. The factors from the supply chain are also missing thus affecting the distribution channels.

## CONCLUSION

The context above proves the potential of precision farming and Intelligent crop planning to achieve the goal of rising food demand along with keeping in mind the environment-related aspects. It can be clearly observed that Information Technology provides some important aspects of farming. The IT-enabled approach provides high yield in farming as well as supports the farmers in various other ways. However, the problem lies in the fact that there is a limitation of data, infrastructure, finances, *etc*, which stops the exploration capacity. The nascent stage of its adoption stops many investments, therefore, creating another big hurdle. The concept is far ahead in developed countries whereas it requires adopting a need-based approach for India. To achieve this aim, a PPP kind of model can be of great help which will help in raising awareness and finances. The development of technical facilities and legal backing to these can further push it in the right direction.

## REFERENCES

[1] "Why farmers today need to take up precision farming", *Down To Earth*. Available From: https://www.downtoearth.org.in/blog/agriculture/why-farmers-today-need-to-take-up-prcision-farming-64659

[2] X. Gramwork, "Precision Farming — Technology infusion in agriculture", *Medium*. Available From: https://gramworkx.medium.com/precision-farming-technology-infusion-in-agriculture-83b72f336b2d

[3] S. Cook, and R. Bramley, "Precision agriculture: Using paddock information to make cropping system internationally competitive", In: *Emerging technologies in agriculture: From ideas to adoption*, 2000.

[4] A. Dobermann, S. Blackmore, S.E. Cook, and V. Adamchuk, "International crop science congress", In: *Precision Farming: Challenges and Future Directions*, 2004.

[5] Massey Ferguson Ltd, "Precision farming systems. A supplement to the cutting edge", *Combine news from Massey Ferguson,* 1995.

[6] B. Basso, L. Sartori, and M. Bertocco, "Precision agriculture - theoretical concepts and practical applications", In: *EdizioniL 'InformatoreAgrarioSpA*, 2005, p. 156.

[7] L. Sartori, and M. Bertocco, "Criteri di scelta per glispandiconcime a distribuzionevariabile (Criteria for selecting spatially variable rate fertiliser spreaders)", *Inf. Agrar.,* pp. 33-38, 2005.

[8] "GPS agriculture: How satellite farming is helping precision agritcch", Available From: https://stories.pinduoduo-global.com/agritech-hub/how-gps-technology-is-helping-prcision-agriculture

[9] R.G.V. Bramley, "Variation in the yield and quality of winegrapes and the effect of soil property variation in two contrasting Australian vineyards", CSIRO Land and Water / Grape and Wine Research and Development Corporation, Final Report on GWRDC Project No. CSL00/01, 2001.

[10] R.G.V. Bramley, *Progress in the development of precision viticulture - Variation in yield, quality and soil properties in contrasting Australian vineyards.* CSIRO Land and Water / Grape and Wine Research and Development Corporation, Final Report on GWRDC Project No. CSL00/01, 2001.

[11] R.G.V. Bramley, "Precision agriculture", CSIRO Land and Water / Grape and Wine Research and Development Corporation, Final Report on GWRDC Project No. CSL02/01, 2002.

[12] R.G.V. Bramley, "Precision viticulture - Tools to optimisewinegrape production in a difficult landscape", CSIRO Land and Water / Grape and Wine Research and Development Corporation, Final Report on GWRDC Project No. CSL02/01, 2002.

[13] A. Castagnoli, and P. Dosso, "Viticolturaassistita da satellite (Satellite aided viticulture)", *Inf. Agrar.,* pp. 77-81, 2001.

[14] B. Basso, L. Sartori, and M. Bertocco, "Precision agriculture - theoretical concepts and practical applications", In: *EdizioniL Informatore Agrario SpA*, 2005, p. 156.

[15] C. Timmermann, R. Gerhards, and W. Kühbauch, *Precis. Agric.,* vol. 4, no. 3, pp. 249-260, 2003. [http://dx.doi.org/10.1023/A:1024988022674]

[16] A. Comparetti, "Precision agriculture: Past, present and future", *International scientific conference on Agriculture Engineering and Environment,* pp. 216-230, 2011.

[17] V.I. Adamchuk, J.W. Hummel, M.T. Morgan, and S.K. Upadhyaya, "On-the-go soil sensors for precision agriculture", *Comput. Electron. Agric.,* vol. 44, no. 1, pp. 71-91, 2004. [http://dx.doi.org/10.1016/j.compag.2004.03.002]

[18] The Hindu BusinessLine, "Towards precision agriculture", *The Hindu Business Line.*

[19] "How is precision agriculture different from smart farming?", Available From: https://www.cropin.com/blogs/how-is-precision-agriculture-different-from-smart-farming

[20] R. K. T. P., "Intelligent farming," *Business Today.* [Online]. Available: https://www.businesstoday.in/magazine/technology-special/story/intelligent-farming-263496-2020-07-08

[21] K. Goel, and A.K. Bindal, "Wireless sensor network in precision agriculture: A survey report", *2018 Fifth International Conference on Parallel, Distributed and Grid Computing (PDGC),* 2018. [http://dx.doi.org/10.1109/PDGC.2018.8745854]

[22] A. Mangla, C. Singh, S. Pal, and A. Bindal, "Deployment of Soil Sensors in WSNs for Precision Agriculture", *International Journal on Future Revolution in Computer Science & Communication Engineering,* vol. 4, no. 3, pp. 146-149, 2018.

[23] N. Zhang, M. Wang, and N. Wang, "Precision agriculture—a worldwide overview", *Comput. Electron. Agric.,* vol. 36, no. 2-3, pp. 113-132, 2002. [http://dx.doi.org/10.1016/S0168-1699(02)00096-0]

[24] A. Balafoutis, B. Beck, S. Fountas, J. Vangeyte, T. Wal, I. Soto, M. Gómez-Barbero, A. Barnes, and V. Eory, "Precision agriculture technologies positively contributing to GHG emissions mitigation, farm productivity and economics", *Sustainability,* vol. 9, no. 8, p. 1339, 2017. [http://dx.doi.org/10.3390/su9081339]

[25] "GIS and Precision Agriculture," *Com.au.* [Online]. Available: https://sugarresearch.com.au/sugar_files/2017/02/IS14015-GIS-and-Precision-Agriculture.pdf

[26] R. Bobby and Grisso, "Precision farming tools: Variable-rate application," *Lib.vt.edu.* [Online].

Available: https://vtechworks.lib.vt.edu/bitstream/handle/10919/47448/442-505_PDF.pdf

[27]   Siddegowda and A. Jayanthila Devi, "A study on the role of Precision Agriculture in Agro-industry", *International Journal of Applied Engineering and Management Letters,* pp. 57-67, 2021.

[28]   J.E. Sawyer, "Concepts of variable rate technology with considerations for fertilizer application", *J. Prod. Agric.,* vol. 7, no. 2, pp. 195-201, 1994.
[http://dx.doi.org/10.2134/jpa1994.0195]

## CHAPTER 8

# Artificial Intelligence and Drones in Smart Farming

**Prabhash Chandra Pathak**[1,*], **Syed Anas Ansar**[1] and **Ajeet Kumar**[2]

[1] *School of Computer Applications, Babu Banarasi Das University, Lucknow, India*

[2] *Mott Macdonald, Bangalore, India*

**Abstract:** Since India is the second-highest populated country in the world and the seventh-largest country in terms of area, which includes hills, plateaus, coastal areas, *etc.*, this situation of land makes a variety of crops and harvest timelines. These timelines seeped into India's culture and festivals. The harvest planning of farmers became a very challenging task due to the variety of land and multitude of harvest timelines as well. To execute this harvest plan, farmers must survey and map their land, but their limited reach restricts them. In view of these restrictions and limitations, drones can be very helpful for farmers; these drones can improve surveying quality and provide a proper harvest timeline as output. Artificial Intelligence-powered drones will give results in three stages: analysis of field planning, tracking the growth and counting of crops, and finally the ripeness tracking and timing of the harvest.

**Keywords:** Artificial intelligence, Drone, Gross domestic product and agriculture.

## INTRODUCTION

The agriculture sector has a long history in India, since the Indus Valley Civilization. After China, India is the world's second-most populous country, where agriculture is the primary source of income and nutrition for 70% of the population [1]. The traditional approach in the agriculture sector is a primitive method of farming that makes extensive use of indigenous knowledge, natural resources, traditional equipment, farmers' cultural values, and organic fertilizer [2]. However, due to the advancement of ICT in the agriculture sector, the working culture has improved over time.

From the invention of the plough to the development of the global positioning system (GPS) to the use of drones to control precise agricultural equipment, technological advancements have shaped agriculture significantly, and humans

---

[*] **Corresponding author Prabhash Chandra Pathak:** School of Computer Applications, Babu Banarasi Das University, Lucknow, India; E-mail: pathakprabhash2@gmail.com

**Praveen Kumar Shukla & Tushar Kanti Bera (Eds.)**

have devised innovative techniques to make farming more efficient and increase crop productivity [3, 4]. According to various types of research, the majority of the Indian population still lives in rural areas; improvement in rural marketing is a common way for the growth of the rural economy [5]. However, due to the advancement of ICT in the agriculture sector, the working culture has improved over time. The agriculture sector is the backbone of the Indian economy, and it plays a very vital role, therefore, it is necessary to discuss how agriculture has contributed to Indian Economy development [6]. The contribution of the agriculture sector may be measured from various aspects such as its share, namely employment generation, Gross Domestic Product (GDP), exports, *etc.* According to Statista, "the industry of agriculture sector employs a large number of people, approximately 60%, which accounts for about 18% of the country's GDP [7]". In India, most of the commercial groups and government plans have begun to work in this area and have seen different approaches in the rural sector, with an emphasis on agriculture [8]. This sector also plays a major role in supporting the industrial sector by providing basic raw materials and food for the industry's personnel. In addition, it generates industrial goods demand; these aspects must be investigated in order to determine the role and importance of agriculture in a given economy. Swarms of locusts have been known to eat plants, crops, and other trees; they have the potential to destroy cultivated crops. This feeding by Locust swarms may cause the shortage of food grains, deprivation, and starvation in societies that rely primarily on the survival of crops. Recently, it invaded different regions of India, including Rajasthan; across 20 regions with 90,000 (approx.) hectares of land devastated [9]. The majority of countries fighting locust swarms rely heavily on organophosphate insecticides. The utilization of technology in the agriculture sector helps to deal with such cases: in Rajasthan, drones have been stationed to ensure that the spraying is done efficiently. Drones can spray insecticides across a 2.5-acre area in under 15 minutes [10]. Drone usage combat locust swarms and provides a secure, immediate solution with this approach. After World War II, the drone concept was applied to coin the word for unmanned aerial vehicles used in target practice. This is a single-purpose mission that involves the destruction of an aircraft. Over the last decade, the meaning of the word has expanded further, and now it's military and civilian unmanned, from remote-controlled toy quad copters to advanced unmanned aerial vehicles used in a variety of commercial applications, including aerial photography. Drones are beneficial for a number of modest agriculture tasks, including crop spraying, soil and field study, irrigation, and day-to-day crop monitoring.

# CONTRIBUTION OF THE AGRICULTURE SECTOR IN DIFFERENT TERMS

## Contribution to Employment

This sector typically employs a huge number of workers as compared to other sectors. Since it is an equally important criterion for accessing the importance of agriculture that includes the population share of the total workforce. In addition, industrialization is increasing day by day in India, therefore, this share may go down in the near future. In India, the number of workers employed in agriculture is still relatively high, notwithstanding an increase in the proportion of workers employed in industry and services [11].

## Contribution to Exports

Another area in which agriculture plays a very important role is the export sector. As we know that nowadays, due to the availability of crop medicines and organic fertilizers, farming produces very high-quality crops, fruits, pulses, *etc.*, and these high-quality products can export to different countries as per their requirement [12].

## Contribution to GDP

It has been noticed that agriculture contributed a significant portion of GDP in most economies prior to industrialization. According to Trading Economics, Agriculture's GDP in India climbed 4076.41 INR Billion to 6630.37 INR Billion from the third quarter to the fourth quarter of 2021 [13]. The agriculture sector in India is projected to account for only about 14% of the country's economy, but it is 42% of total employment [14]. For the least developed countries, agriculture acts as the heart of economic growth. The agriculture sector is an important pillar of the economy as it supplies the majority of essential foods, which range from 25 to 95%. This sector accounts for a considerable amount of GDP, ranging from 30 to 60%, and employs 40 to 90% of the labour force in the majority of situations. According to the 2020-2021 Economic Survey, the contribution of the agriculture sector to GDP has risen for the first time, nearly 20% in 17 years, as shown in Fig. (**1**), [15, 16]. All of this will be feasible due to the utilization of modern technology in agriculture.

As the accelerations in the industrialization process, the non-agricultural sectors raise the GDP share, and at the same time, agriculture's proportional share of the economy falls behind industry and services. It simply means that industrial and service sector production grows faster than agricultural production. The transformation in the economy's structure is because of more industrialization and advancements in this sector. The farms' productivity is very important for many reasons, for more food, increasing market growth, migration of labourers, and citizens' income. The efficient distribution of scarce resources can also be increased through farming productivity, as improving productivity is always an integral part of productive and effective farming. Nowadays, innovative technologies and techniques have offered farmers the opportunity to boost agricultural productivity and long-term sustainability [17]. In this paper, researchers have discussed key factors to provide knowledge on how to increase agricultural productivity.

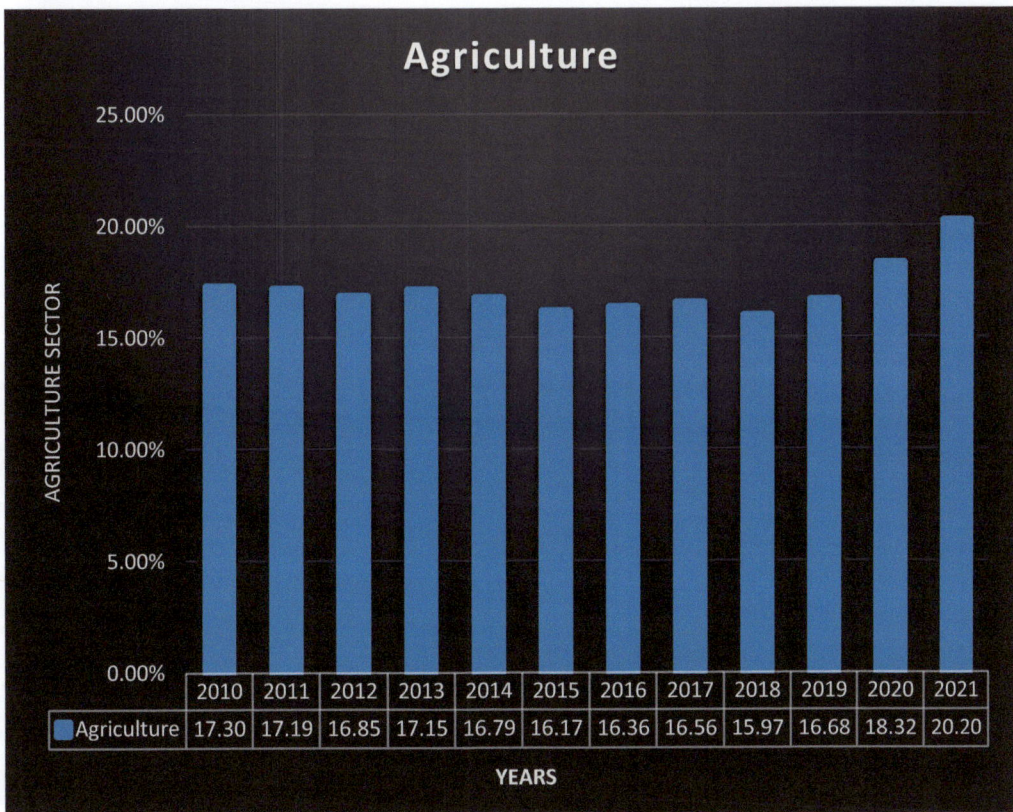

**Agriculture**

| | 2010 | 2011 | 2012 | 2013 | 2014 | 2015 | 2016 | 2017 | 2018 | 2019 | 2020 | 2021 |
|---|---|---|---|---|---|---|---|---|---|---|---|---|
| Agriculture | 17.30 | 17.19 | 16.85 | 17.15 | 16.79 | 16.17 | 16.36 | 16.56 | 15.97 | 16.68 | 18.32 | 20.20 |

**Fig. (1).** Contribution of agriculture towards India's GDP.

## METHODS TO IMPROVE FARMING PRODUCTIVITY

Enhancement in farming production is the main demand and need of every farmer. There are numerous elements that can enhance agricultural productivity. This paper discusses several techniques for increasing agricultural production.

## Reformation of Land

For the improvement of productivity, reformation of the land is the first and predominant method. Machines and new technologies like drones can be used to improve the productivity of the land [18]. Machines and emerging technologies have the qualities to identify and make rugged farming lands smooth so that farmers can work efficiently in the field. The most effective strategy to boost farming work and increase productivity is through land reforms.

## Challenges

The main problem to implement this method for the increased productivity of land is the identification of land areas where reformation of land is required.

## Inter-plantation

Inter-plantation is a most acceptable practice that involves planting multiple crops simultaneously. This method is the most cost-effective technique to increase the production of your land.

## Challenges

Some crops work well together for inter-planting, while others do not. In addition, proper plant monitoring is essential on a regular basis to determine the suitability of inter-planting crops.

## Smart Water Management

Plant watering is a fundamental aspect of the planting process, and water management can improve crop output. One can enhance the yield by up to 50% by using a sprinkler watering system [19]. In addition, for a better irrigation system, manufacturing canals, *i.e.*, tube wells, is a good option for crop safety.

## Challenges

The main challenge in the watering of plants is to maintain the moisture up to a required level as per the need of plants. Moisture measurement is the big challenge for enhancement of productivity of the land.

## Heat Tolerant Varieties

To maintain the yield of the plant at high temperatures, heat-tolerant varieties are very important. If we improve the heat-tolerant varieties of the land, then crop yield can be increased by up to 23% [20].

## Challenges

To measure heat tolerance of the plant is one of the major issues to maintain the productivity of the land and yield of the crops.

## Plant Protection

According to the majority of farming scientists, approximately 5% of crops are damaged by pests, insects, and diseases [21]. The vast majority of farmers rely on recently developed pesticides and medicines. This development helped in crops production as well as yields of the crops. Farmers must use these medicines.

## Challenges

To keep the plant protected from insects and diseases is a big challenge for farmers but the time and quantity of spray of insecticides and pesticides are the most significant problems.

# USE OF TECHNOLOGY IN AGRICULTURE TO OVERCOME CHALLENGES

There are many areas in which technology affects agriculture, such as seed technology, fertilizers, pesticides, *etc*. Pest resistance has resulted in the growth of crop yields through genetic engineering and biotechnology. Harvesting, tilling, and physical work have all become more efficient with the help of mechanization. Technology has improved irrigation methods and transportation infrastructures, and the effect can be observed in all areas [22]. Artificial intelligence, robotics, blockchain technology, precision agriculture, and other new technologies are among them. The following are some of the most significant technological advancements in agriculture:

## Improvement in Productivity Through the Mechanization of Agriculture

Combine harvesters are becoming more popular due to their ability to reduce manual labour and enhance harvesting procedures. Asin India, agriculture is characterized by small landholdings, necessitating collaboration with others in order to benefit from modern machinery.

## Climate Forecasting Prediction Through Artificial Intelligence

Artificial intelligence (AI) has made a significant contribution to agriculture. Data collection and informed decision-making are aided by advanced AI-based tools and technology [23]. It facilitates different services such as forecasting weather reports 24/7, with important information on humidity, temperature, soil, rainfall, other variables, remote sensors, drones, satellites, *etc.* Furthermore, in a country like India, AI is difficult to catch on due to marginal farming, fragmented landholdings, and other factors, and the adoption of new technology is considerably slower than it should be. However, there is little doubt that AI-based technology can bring accuracy to large-scale farming and increase production exponentially.

### Improving Farm Yields and Supply Chain Management Uses Big Data.

In smart farming, big data plays a major role and it provides benefits across the entire market and the supply chain. The data collection, as well as its compilation and subsequent processing, make it extremely valuable for problem-solving strategies and decision-making, as well as broadening the scope of how big data works [24]. Nowadays, agriculture is becoming more complex, and it is dependent on a wide range of factors. The increased acquisition of data, as well as the usage of complex data, necessitates more meaningful interpretation and management. Data can be gathered from a variety of external sources, including markets, social media, sensor/machine, and supplier networks.

### *Why Agricultural Drone Should be adopted?*

The drone's diversity has received the majority of the industry's recognition. This technology is very useful to consider for the future of the agrarian community.

### *How can Drones Support Indian Agriculture?*

Drones can help farmers address a variety of issues, not only improving overall performance but also encouraging them to overcome other obstacles and reap several benefits through precision agriculture. The UAVs (unmanned aerial vehicles)/drones fill the gap that occurred due to human error, inefficiency, and conventional methods in farming. From 2019 to 2025, the global agriculture drone market is predicted to increase at a CAGR of 31.1%, reaching $5.19 billion [25]. Implementing this technology aims to eliminate any uncertainty or guesswork and instead focus on reliable and accurate data. Weather, soil conditions, and temperature are all important aspects of agriculture. The agriculture drones enable farmers to adapt to individual circumstances and make thoughtful decisions as a result. Crop treatment, crop health, crop scouting, field soil analysis, irrigation, and crop damage assessments are all aided by the information gathered. The drone survey increases crop yields while reducing time and costs. Experts predict that by 2050, the population of the world will have reached 9 billion. Agricultural consumption is also expected to rise by about 70% at the same time [26]. Drone technology integrates different technologies, mainly machine learning (ML), artificial intelligence (AI), and remote sensing capabilities that are increasingly popular due to their benefits. With its online 'Digital Sky Platform' the central government has acknowledged the importance of machine learning, artificial intelligence, and unmanned aerial vehicles (UAVs). India's drone start-ups have taken advantage of the opportunity to improve their technological capabilities.

## WORKING OF DRONE TECHNOLOGY

After a thorough understanding of drone characteristics, one can understand farm drones. Drones typically contain a navigation system, GPS, several sensors, high-resolution cameras, programmable controllers, and autonomous drone tools. The DJI is one of the most well-known drones in the market. The majority of farmers now use satellite imagery as a starting point for farm management. Unmanned aerial vehicles (UAVs) equipped with modern technology can obtain more exact data for precision agriculture than satellites. They then input the data into agri-tech tools to generate useful information.

The following stages are involved in capturing data from an agriculture drone [27]:

**Area Analysis:** The first phase involves the territory identification that will be tested; it basically analyses the region, results defining a boundary, and ultimately, maintains the technical GPS data by uploading it into the navigation system of drones.

**Using Autonomous Drones:** As UAVs are not dependent, they enter flight patterns into their pre-existing data collection system.

Uploading the data: Once all necessary data has been captured using sensors, namely the RGB/multispectral sensor. This data is processed through different software for further interpretation and analysis.

**Output:** In this stage, after data collection, it is formatted in such a manner that it can be easily understandable by farmers. This phase, brings them one step closer to precision farming, the prominent approaches are photogrammetry and 3D mapping for displaying large amounts of data.

## BEST DRONE PRACTICES

Drone technology swiftly restores traditional farming methods, which are then carried out as follows [27, 28].

**Irrigation Monitoring:** Drones with hyperspectral, thermal, or multispectral sensors detect areas that are too dry or require the farmer's attention (Fig. **2**).

Irrigation monitoring yields calculations of the vegetation index to help realize the health of crops and emitted heat/energy. Drone survey helps improve water efficiency and disclose potential pooling/leaks in irrigation by providing Irrigation monitoring and yield calculations of the vegetation index to help realize the crop's health and emitted heat/energy.

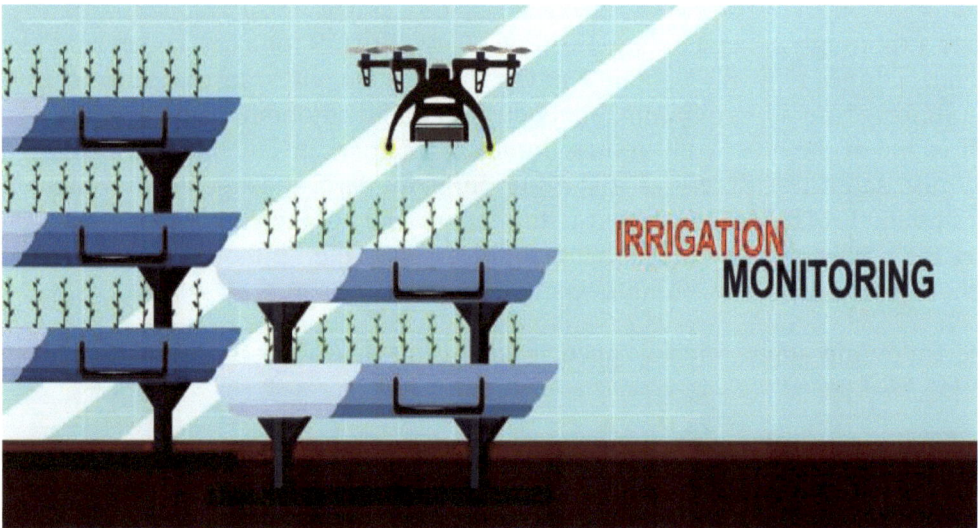

**Fig. (2).** Irrigation Monitoring (https://ik.imagekit.io/equinoxsdrones/blog/img/importance_of_drone_ technology-in-Indian-agriculture-farming/IRRIGATION_MONITORING_m_4WhiIWE.jpg).

**Monitoring and Surveillance of the Crop Health:** Tracking the health of the vegetation and spotting bacterial/fungal diseases early on is critical. Plants that reflect various quantities of green light and Near-infrared spectroscopy (NIRS) light can be identified by agriculture drones. This information is used to create multispectral images that can be used to track crop health. Crops can be saved if they are monitored closely, and any faults are discovered quickly. In the event of crop failure, the farmer can document the losses in order to file proper insurance claims.

**Field Soil Analysis:** Farmers can use the drone survey to learn more about the soil conditions on their land. Multispectral sensors capture data that can be used for field soil analysis, seed planting patterns, nitrogen management, and irrigation. Farmers can thoroughly examine their soil conditions using precise photo-grammetry/ 3D mapping.

**Crop Damage Assessment:** The agricultural drones equipped with RGB sensors and multispectral can also identify weeds, diseases, and pests in farm regions Fig. (3). This provides the exact amount of chemicals required to combat these infestations that are known as a result of this research, which reduce the farmer's costs.

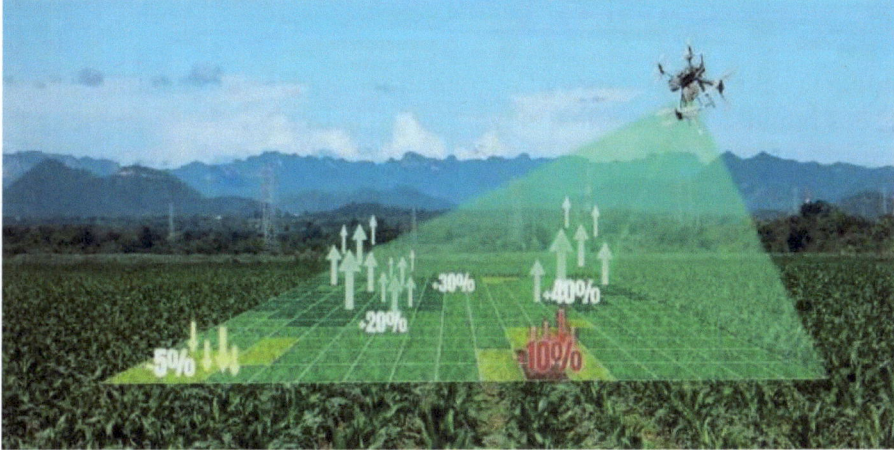

**Fig. (3).** Crop Damage Assessment (Source: https://www.vertica.com/blog/dx-radio-episode-3-climate-corporation-and-digital-agriculture).

**Agricultural Spraying:** Human interaction with such dangerous substances is restricted because of drone agricultural spraying. Agri-drones can complete this duty considerably faster than cars or planes Fig. (**4**). Drones equipped with RGB and multispectral sensors can pinpoint problem areas and address them effectively. According to experts, aerial spraying with drones is five times faster than previous approaches.

**Fig. (4).** Pesticides spraying (Source: https://www.agroberichtenbuitenland.nl/actueel/nieuws/2021/05/21/hungary-quick-news-ksh).

**Planting:** Indian businesses have developed Drone-planting systems that facilitate drones to shoot seeds, pods, and essential nutrients into the soil. Not only does this technique cut expenses by over 85%, but it also improves consistency and efficiency.

**Livestock Tracking:** Farmers can use the drone survey to keep track of not only their crops but also their cattle's movements. Thermal sensor technology aids in the recovery of missing animals and the detection of injury or illness. Drones are capable of performing this duty well, and this contributes significantly to the development of vegetation.

## BENEFITS OF DRONE TECHNOLOGY

As new technologies are introduced by innovators, their commercial applications grow by the day. Drone usage limits have been eased by the government, and start-ups are being encouraged to come up with innovative concepts. Drone surveys are becoming more cost-effective as they become more widespread. They offer numerous advantages in agriculture; few are discussed below [29, 30]:

**Enhanced Production:** Comprehensive irrigation planning, sufficient crop health monitoring, better soil health knowledge, and response to environmental changes are all ways for farmers to improve their output capacities.

**Effective and adaptive Ttchniques:** Drone use allows farmers to receive regular information on their crops and aids in the development of more effective farming techniques. They can adjust to changing weather conditions and utilize resources efficiently.

**Greater safety of farmers:** Using drones to spray pesticides in difficult-to-reach terrains, taller crops, contaminated areas, and a power line is safer and more convenient for farmers. It also aids farmers in avoiding crop spraying, resulting in less pollution with less usage of chemicals in the soil.

**10x faster data for quick decision-making:** The survey data of drones provides reliable data to farmers and allows them to make rapid and deliberate judgments without second-guessing, saving time spent on crop scouting. The drone's various sensors allow it to capture and analyze data from the entire field. The information can be used to target issue regions such as sick or unhealthy crops, various coloured crops, moisture levels, and so on. Even varieties of sensors are installed for different crops on the drone, and it provides a more precise and versatile management system for crops.

**Less wastage of resources:** Agri-drones allow for the most efficient use of all resources, including water, pesticides, fertilizer, and seeds.

**99% Accuracy rate:** The drone survey assists farmers in calculating the exact size of land, soil mapping, and segmenting different crops.

**Useful for Insurance claims:** Farmers utilize the data collected by drones to file crop insurance claims in the event of damage. While being insured, they even analyze the risks and losses linked with the land.

**Evidence for insurance companies:** Agri-drones are used by agricultural insurance companies for efficient and reliable data. They document the losses that have happened in order to calculate the appropriate monetary compensation for the farmers.

## DISCUSSION

The agriculture sector is the primary source of income, especially in rural areas as 70% of the country's population depends on it. In 2019-2020, at a rate of 2.1 percent, the agriculture sector was rising in parallel with other industries in 2019-2020. In this sector, adopting new approaches to technology enhances agriculture productivity. Despite several advances in the agriculture sector, the service sector is declared to be the most substantial contributor to India's GDP Fig. (**5**). The graph below shows the worldwide record of several sectors and this statistic demonstrates agriculture's contribution (*i.e.*, 4.35% approximately), which is relatively lower than that of other sectors [31].

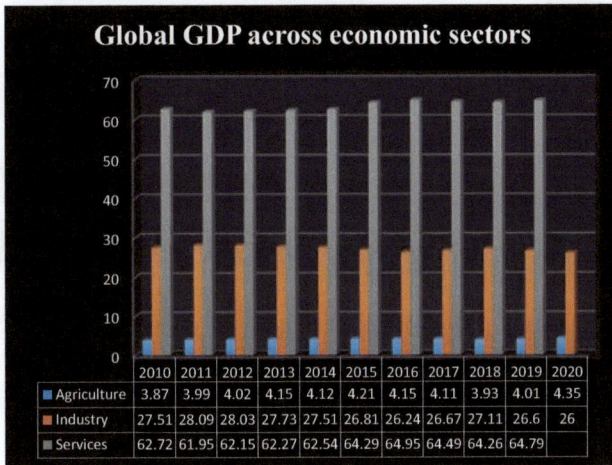

**Global GDP across economic sectors**

| | 2010 | 2011 | 2012 | 2013 | 2014 | 2015 | 2016 | 2017 | 2018 | 2019 | 2020 |
|---|---|---|---|---|---|---|---|---|---|---|---|
| ■ Agriculture | 3.87 | 3.99 | 4.02 | 4.15 | 4.12 | 4.21 | 4.15 | 4.11 | 3.93 | 4.01 | 4.35 |
| ■ Industry | 27.51 | 28.09 | 28.03 | 27.73 | 27.51 | 26.81 | 26.24 | 26.67 | 27.11 | 26.6 | 26 |
| ■ Services | 62.72 | 61.95 | 62.15 | 62.27 | 62.54 | 64.29 | 64.95 | 64.49 | 64.26 | 64.79 | |

**Fig. (5).** Global GDP across economic sectors.

India is the world's leading producer of milk, pulses, wheat, sugarcane, rice, spices, and other agricultural products. These activities in the agro sector also contribute a significant value to the economy. The agriculture sector in India accounts for 20.2% of the country's GDP (Gross Domestic Product). However, when compared to the industry and services sectors' contributions to GDP, agriculture's contribution is minimal, as shown in the graph below [15]:

According to experts, it is quite challenging to grasp this technology at first, but once it is learned, it will produce a result in a short time and can boost agricultural GDP by 1-1.5 percent [32]. As the researchers of this paper previously stated, "In India, drone technology is the future of the Agrarian community". This technology has the potential to change the traditional farming process in innumerable ways. Even if only 10-15% of this technology is used in agriculture, it will have a significant impact on the sector. Drone facilitates different small agriculture applications such as crop spraying, soil and field analysis, irrigation, and day-to-day crop monitoring. It is 40-60% faster than the conventional spraying approach as compared to manual spraying and saves 30 to 40% on chemicals as well as preserving 90% of the water [33]. All of these potential benefits of drones in agriculture led to a surge in the Indian economy's GDP Fig. (6).

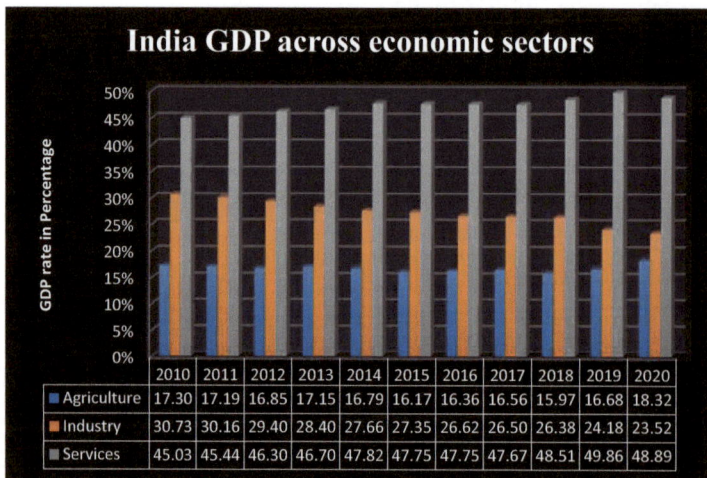

**India GDP across economic sectors**

| | 2010 | 2011 | 2012 | 2013 | 2014 | 2015 | 2016 | 2017 | 2018 | 2019 | 2020 |
|---|---|---|---|---|---|---|---|---|---|---|---|
| ■ Agriculture | 17.30 | 17.19 | 16.85 | 17.15 | 16.79 | 16.17 | 16.36 | 16.56 | 15.97 | 16.68 | 18.32 |
| ■ Industry | 30.73 | 30.16 | 29.40 | 28.40 | 27.66 | 27.35 | 26.62 | 26.50 | 26.38 | 24.18 | 23.52 |
| ■ Services | 45.03 | 45.44 | 46.30 | 46.70 | 47.82 | 47.75 | 47.75 | 47.67 | 48.51 | 49.86 | 48.89 |

**Fig. (6).** India GDP across economic sectors.

## CONCLUSION

The agriculture sector is a major driver of the economy, as it provides the bulk of basic foods. This is a crucial component of smart development, but due to a lack of technology, its contribution is minimal to the Indian Economy GDP. Among the 226 countries, agriculture is the most important economic sector in nine of them. The agricultural industry accounts for 60.7 percent of Sierra Leone's GDP. The agricultural sector accounts for more than half of the GDP in the three countries. In addition, India is placed 40th in the world, with agriculture accounting for 20% of GDP. This study primarily explored issues in the agriculture industry that can be remedied through the use of drones in agriculture. If this technology is used effectively in agriculture, it will increase GDP percentage and contribute a higher percentage of GDP than other sectors. As a result of the integration of modern technology with farming, agricultural production has increased, including GDP growth in the agriculture sector. This will have a significant positive impact on India's economy. In the future, researchers of this study will provide a roadmap to check the fertility of the soil with the help of drones.

## REFERENCES

[1]　M. Banoo, B.K. Sinha, G. Chand, S. Dogra, and R. Sinha, "Traditional Agriculture: Alternative Practices for Climate Change Mitigation", *Agricolation,* vol. 1, no. 5, pp. 14-18, 2021.

[2]　R. Singh, and G.S. Singh, "Traditional agriculture: A climate-smart approach for sustainable food production", *Energy Ecol. Environ.,* vol. 2, no. 5, pp. 296-316, 2017. [http://dx.doi.org/10.1007/s40974-017-0074-7]

[3]　J.V. Rane, "A Study on E-Agriculture and Rural Development in Gondiya District of Maharashtra (India)", *J. Emerg. Technol. Innov. Res.,* vol. 9, no. 1, pp. f1-f5, 2022. [JETIR].

[4]　S. Kumar, *Educ. Technol.,* vol. 11, no. 1, pp. 59-80, 2022.

[5]　G.V. Bhasker, "Agriculture Role of Indian Economy", *International Journal of Trend in Scientific Research and Development,* vol. 1, no. -6, pp. 1066-1067, 2017. [http://dx.doi.org/10.31142/ijtsrd5790]

[6]　J. Ahmad, D. Alam, and M.S. Haseen, "Impact of climate change on agriculture and food security in India", *Int. J. Agric. Environ. Biotechnol.,* vol. 4, no. 2, pp. 129-137, 2011.

[7]　Available From: https://www.statista.com/topics/4868/agricultural-sector-in-india/#dossierKeyfigures (Last Accessed: 8 Jan 2022)

[8]　S. Sharma, and J. Singh, "Farmer Demography and Use of Information and Communication Technology for Agriculture in India", *Empirical Economics Letters,* vol. 20, no. 5, pp. 51-57, 2021.

[9]　T.W.C. Explainer, Why This Year's Locust Attack in India is the Worst in Recent History. Available From: https://weather.com/en-IN/india/news/news/2020-05-29-twc-explainer-why-this-year-locust-attack-india-worst-recent-history (Last Accessed on : 8 Jan 2022).

[10]　S.C. Alvarez-Ventura, *Measuring Impacts of Neem Oil and Amitraz on Varroa destructor and Apis Mellifera in Different Agricultural Systems of South Florida.* FIU Electronic Theses and Dissertations, 2011, p. 490. [http://dx.doi.org/10.25148/etd.FI11120503]

[11]　"India: Issues and Priorities for Agriculture". Available From: https://www.worldbank. org/en/news/feature/2012/05/17/india-agriculture-issues-priorities (Last Accessed on: 8 Jan, 2022).

[12]　C.L. Thomas, G.E. Acquah, A.P. Whitmore, S.P. McGrath, and S.M. Haefele, "The effect of different organic fertilizers on yield and soil and crop nutrient concentrations", *Agronomy (Basel),* vol. 9, no. 12, p. 776, 2019. [http://dx.doi.org/10.3390/agronomy9120776]

[13]   "India GDP From Agriculture", Available From: https://tradingeconomics.com/india/gdp-fro-
       -agriculture (Accessed on: 9 Jan 2022).

[14]   "India: Distribution of the workforce across economic sectors from 2009 to 2019", Available From:
       https://www.statista.com/statistics/271320/distribution-of-the-workforce-across-economic-sec-
       ors-in-india/

[15]   Aaron O'Neill, "Distribution of gross domestic product (GDP) across economic sectors in India 2020",
       Online available at: India - Distribution of gross domestic product (GDP) across economic sectors
       2020 | Statista, Last Accessed on: 9 Jan 2022.

[16]   "Contribution of Agriculture Sector towards GDP Agriculture has been the bright spot in the Economy
       despite COVID-19", Available From: https://pib.gov.in/PressReleaseIframePage.aspx?PRID=1741942

[17]   K. Takahashi, R. Muraoka, and K. Otsuka, "Technology adoption, impact, and extension in developing
       countries' agriculture: A review of the recent literature", *Agricultural Economics,* vol. 51, no. 1, pp.
       31-45, 2020.

[18]   H.S. Abdullahi, F. Mahieddine, and R.E. Sheriff, "Technology impact on agricultural productivity: A
       review of precision agriculture using unmanned aerial vehicles", *International conference on wireless
       and satellite systems,* pp. 388-400, 2015.
       [http://dx.doi.org/10.1007/978-3-319-25479-1_29]

[19]   S.K. Biswas, A.R. Akanda, M.S. Rahman, and M.A. Hossain, "Effect of drip irrigation and mulching
       on yield, water-use efficiency and economics of tomato", *Plant, Soil and Environment,* vol. 61, no. 3,
       pp. 97-102, 2015.
       [http://dx.doi.org/10.17221/804/2014-PSE]

[20]   M.U. Hassan, M.U. Chattha, I. Khan, M.B. Chattha, L. Barbanti, M. Aamer, and M.T. Aslam, "Heat
       stress in cultivated plants: Nature, impact, mechanisms, and mitigation strategies—A review", *Plant
       Biosystems-An International Journal Dealing with all Aspects of Plant Biology,* vol. 155, no. 2, pp.
       211-234, 2021.

[21]   K. Thenmozhi, and U. Srinivasulu Reddy, "Crop pest classification based on deep convolutional
       neural network and transfer learning", *Comput. Electron. Agric.,* vol. 164, p. 104906, 2019.
       [http://dx.doi.org/10.1016/j.compag.2019.104906]

[22]   A. King, "Technology: The future of agriculture", *Nature,* vol. 544, no. 7651, pp. S21-S23, 2017.
       [http://dx.doi.org/10.1038/544S21a] [PMID: 28445450]

[23]   N.N. Misra, Y. Dixit, A. Al-Mallahi, M.S. Bhullar, R. Upadhyay, and A. Martynenko, "IoT, big data
       and artificial intelligence in agriculture and food industry", *IEEE Internet Things J.,* 2020.

[24]   M.S. Farooq, S. Riaz, A. Abid, K. Abid, and M.A. Naeem, "A Survey on the Role of IoT in
       Agriculture for the Implementation of Smart Farming", *IEEE Access,* vol. 7, pp. 156237-156271,
       2019.
       [http://dx.doi.org/10.1109/ACCESS.2019.2949703]

[25]   "Top 10 companies in agriculture drone market", Online available at: Top 10 companies in agriculture
       drone market (meticulousblog.org), Last accessed on: 10 Jan 2022.

[26]   "How to Feed the World in 2050", Online available at: Microsoft Word - Synthesis_Report.doc
       (fao.org), Last Accessed on: 10 Jan 2022.

[27]   A. Rani, A. Chaudhary, N.K. Sinha, M. Mohanty, and R. Chaudhary, "Drone: The green technology
       for future agriculture", *Harit Dhara,* vol. 2, no. 1, pp. 3-6, 2019.

[28]   D. van der Merwe, D.R. Burchfield, T.D. Witt, K.P. Price, and A. Sharda, "Drones in agriculture",
       *Adv. Agron.,* vol. 162, pp. 1-30, 2020.
       [http://dx.doi.org/10.1016/bs.agron.2020.03.001]

[29]   M. Reinecke, and T. Prinsloo, "The influence of drone monitoring on crop health and harvest size", *1st*

*International Conference on Next Generation Computing Applications (NextComp),* pp. 5-10, 2017. [http://dx.doi.org/10.1109/NEXTCOMP.2017.8016168]

[30]   M. Reinecke, and T. Prinsloo, "The influence of drone monitoring on crop health and harvest size", *2017 1st International conference on next generation computing applications (NextComp),* pp. 5-10, 2017.

[31]   "Share of economic sectors in the global gross domestic product (GDP) from 2010 to 2020", Available From: https://www.statista.com/statistics/256563/share-of-economic-sectors-in-the-global-gross-domestic-product/

[32]   "Drone usage to boost agri GDP by 1-1.5%: Experts", Online available at: Drone usage to boost agri GDP by 1-1.5%: Experts - The Economic Times (indiatimes.com), Last accessed on: 10 Jan, 2022.

[33]   "Drones in Agriculture: A booming sector?", Online available at: Drones in Agriculture: A booming sector? - Maxellco, Last accessed on: 11 Jan 2022.

# SUBJECT INDEX

## A

Absorption, nutrient 1
Acids, fatty 2
Adaptive neural-fuzzy inference system 44
Adolescent mayflies 5
Agricultural 2, 3, 136, 142
  consumption 136
  products 2, 3, 142
Agriculture 1, 2, 9, 53, 54, 55, 114, 119, 120,
    130, 131, 132, 135, 136, 142, 143
  accounting 143
  applications 142
  industry 143
  sustainable 119
Algorithm 10, 12, 13, 14, 17, 19, 20, 72, 73,
    78, 83, 99, 100, 101
  back-propagation 99
  propagation 99
ANN, structure of 98
Architecture, neural community 54
Artificial neural networks (ANN) 1, 2, 4, 5,
    10, 15, 17, 19, 20, 21, 25, 98, 99, 101
Augmentation methods 48
Automatic threshold technique 12
Automating time consumption 70
Automation manufacturing systems 57
Autonomous control system 121

## B

Backpropagation algorithm 99, 106
Balaenoptera 5
Big data analysis 55
Blast 26, 47
  illnesses 26
  pearl millet diseases 47

## C

CNN 26, 46, 50

algorithm 50
architecture 26, 46, 50
Computer-based method 43
Computer vision 3, 24, 42, 43, 44, 53, 69, 70,
    71, 81
  framework 44
  learning-enabled 70
  systems 24, 44, 53
  tasks 43, 81
Conditions 3, 55, 56, 62, 70, 115, 122
  atmospheric 3
  climatic 55, 62
  stress 122
Convolutional neural networks 3, 81
Crop(s) 24, 42, 43, 54, 56, 100, 116, 117, 119,
    120, 123, 126, 131, 132, 135
  disease recognition 24
  diseases 43
  management 100, 123
  measurements 120
  medicines 132
  nutrition 56
  plantation 54
  production 42, 116, 117, 119, 126, 135
  productivity 116, 131
  protection 56
Crop damage 53, 136, 138, 139
  assessment 136, 138, 139
Crystalline silicon 63

## D

Data 100
  filtering methods 100
  mining algorithms 100
Data analysis 107, 124
  post-processing 124
DC 62, 63, 69
  devices 63
  electricity 62
  load 63
  motors 69

www.ingramcontent.com/pod-product-compliance
Lightning Source LLC
Chambersburg PA
CBHW041708210326
41598CB00007B/582